RIPARIAN FORESTS IN CALIFORNIA

THEIR ECOLOGY AND CONSERVATION

A Symposium Sponsored by

Institute of Ecology
University of California
Davis, California

and

Davis Audubon Society

May 14, 1977

Edited by Anne Sands

Originally published as Institute of Ecology Publication No. 15.

Printed in the United States of America.

No. 1   CALIFORNIA ASSEMBLY COMMITTEE ON NATURAL RESOURCES, PLANNING AND PUBLIC WORKS, "Man's Effect on California Watersheds," Report by the Institute of Ecology, University of California, Davis (Sacramento, Calif., January, 1967).

No. 2   GOLDMAN, C. R. (ed.), "Environmental Quality and Water Development," Report to the National Water Commission (Davis, Calif.: Institute of Ecology, University of California, 1971). (Not available).

GOLDMAN, C. R., J. McEvoy III and P. J. Richerson (eds.), "Environmental Quality and Water Development," Revised version, (San Francisco: W. H. Freeman and Co., 1973).

No. 3   CALIFORNIA ASSEMBLY COMMITTEE ON NATURAL RESOURCES AND CONSERVATION, "Public Policy for California Forest Lands," Report by the Institute of Ecology, University of California, Davis (Sacramento, Calif., April, 1973). (Not available).

No. 4   STEBBINS, G. L. and D. W. Taylor, "A Survey of the Natural History of the South Pacific Border Region, California," Report to the National Park Service, U. S. Dept. of the Interior, (Davis, Calif.: Institute of Ecology, University of California, 1973). (Not available).

No. 5   LOVE, R. M. et al., "Maintaining the Environmental Quality of California Wildlands," (Davis, Calif.: Institute of Ecology, University of California, 1973).

No. 6   LOVE, R. M. (ed.), "Resources for the Future," (Davis, Calif.: Institute of Ecology, University of California, 1974).

No. 7   LOVE, R. M. (ed.), "The California Annual Grassland Ecosystem," (Davis, Calif.: Institute of Ecology, University of California, 1975). Second Printing, 1976.

No. 8   LOVE, R. M. and T. E. Adams, Jr. (eds.), "A New Look at California's Brushland Resource," (Davis, Calif.: Institute of Ecology, University of California, 1976).

No. 9   JAIN, S. K. (ed.), "Vernal Pools: Their Ecology and Conservation," (Davis, Calif.: Institute of Ecology, University of California, 1976).

No. 10   FOIN, JR., T. (ed.), "Visitor Impacts on National Parks: The Yosemite Ecological Impact Study," (Davis, Calif.: Institute of Ecology, University of California, 1977).

No. 11   THORPE, Linda J., "Land Use Mapping Programs User's Manual," (Davis, Calif.: Institute of Ecology, University of California, 1977).

No. 12   WINKLER, David W., (ed.), "An Ecological Study of Mono Lake, California," (Davis, Calif.: Institute of Ecology, University of California, 1977).

No. 13   ORLOVE, Benjamin S., "Cultural Ecology: A Critical Essay and a Bibliography," (Davis, Calif.: Institute of Ecology, University of California, 1977).

No. 14   RICHERSON, Peter J., C. Widmer and T. Kittel, "The Limnology of Lake Titicaca (Peru-Bolivia), A Large, High Altitude Tropical Lake," (Davis, Calif.: Institute of Ecology, University of California, 1977).

No. 15   SANDS, Anne (ed.), "Riparian Forests in California: Their Ecology and Conservation," (Davis, Calif.: Institute of Ecology, University of California, 1977).

No. 25   GROTH, A. J. and H. G. Schutz, "Voter Attitudes on the 1976 Nuclear Initiative," (Environmental Quality Series) (Davis, Calif.: Institute of Ecology and Institute of Governmental Affairs, University of California, November, 1976).

TABLE OF CONTENTS

* Indicates edited versions of oral presentations.

FOREWORD

These proceedings are the result of a second symposium on Riparian Forests in California held at the Davis Campus of the University of California on May 14, 1977. The conference was co-sponsored by the Davis Audubon Society and the U. C. Davis Institute of Ecology. Our goals were to expand the knowledge of California's riverine ecosystems and to stimulate public interest in their conservation.

The symposium topics were divided into two sessions: one dealing with historical and ecological subjects and the other with management and preservation. Papers were arranged so that background data came first and were followed by a developmental sequence beginning with the evolution of riparian ecosystems and ending with a discussion of the value of riparian forests in today's society. A lively discussion between speakers and members of the audience lasted well beyond the adjournment schedule. Selections from that interaction are summarized at the end of these proceedings.

On May 22, 1976, the first Riparian Forest Conference was held in Chico, California. That meeting was sponsored by the Davis and Altacal Audubon Societies, and organized by David Gaines. Several speakers at that session, notably Kenneth Thompson, Felix Smith, James Burns and Daniel Frost, were invited to return and present "progress reports" on the status of riparian ecosystems at the 1977 symposium. Much of what they reported was discouraging, and their comments reinforced the urgency of the riparian forest situation we now face in California.

However, the energy and enthusiasm of the speakers and their audience of nearly 200 were inspiring. By the end of a long day of excellent presentations, there was a unanimous call for citizen action which resulted in the conception of the Riverlands Council. This group will act to influence state legislation and to stimulate local controls over riparian forest management through public education. We anticipate that future symposia will be designed as workshops to educate and train participants so they may initiate conservation efforts in their own communities. Public awareness must now be coupled with political pressure. Governmental agencies which have jurisdiction over the fate of riverbanks must be made aware of the significance and uniqueness of these ecosystems. We must study these agencies' surveys, participate in their hearings, join their advisory committees and become well armed with facts and determination. However, even federal and state agencies have restrictions on their spheres of influence. Almost 95% of the yet unspoiled remnants of riparian hardwoods in California are in private ownership. Each year more of these areas are bulldozed for orchards, cut for pulpwood and cleared for "stream bank protection."

Several approaches can be made to solve the riparian protection problem. Land use plans must be established at county and state levels to encourage recreational and open space easements as well as wildlife sanctuaries. Zoning laws should be altered to relieve land owners from heavy taxes on riparian forest (many farmers are taxed on their forests as if they were fruit orchards). Forestry management acts should be amended to protect riparian species. Private landowners should be offered reasonable alternatives to tree cutting, such as tax deductible donations of land to non-profit, private organizations like the American Land Trust, Audubon, and the Nature Conservancy. Prime riverine forest land should be purchased by conservation groups if all other measures fail. We must all publicize what we know about the Riparian Forests and work together to bring about the political changes necessary to preserve these very special and vulnerable ecosystems.

Anne Sands

RIPARIAN FORESTS
THEIR ECOLOGY AND CONSERVATION

PROGRAM

May 14, 1977

Kleiber Hall
U. C. Davis

8:30   Registration and Refreshments
9:00   Anne Sands -- Introduction to the Ecology Session
9:05   Felix Smith -- Historical Review of Riparian Forests
9:20   Warren Roberts -- The Uniqueness of Riparian Ecosystems
9:40   Robert Robichaux -- Evolution of California's Riparian Forests
9:50   Kenneth Thompson -- The Pristine Riparian Forests
10:10   Edward Keller -- The Fluvial Systems: Selected Observations
10:35   Refreshment Break
10:45   Susan Conard -- Riparian Vegetation and Flora of the Sacramento Valley
11:00   David Gaines -- The Valley Riparian Forests of California: Their Importance
                    to Bird Populations
11:20   Donald Alley -- Habitats of Native Fishes in the Sacramento River Basin

NOON   Lunch Break

1:15   Geoffrey A. Wandesforde Smith -- Introduction to the Conservation Session
1:20   John F. Dunlap -- Recent Legislation to Study and Protect California's Riparian
                    Forests
1:40   Fred Kindel -- Environmental Applications in Corps of Engineer's Work
2:05   James Burns -- The Upper Sacramento River Task Force: A Progress Report
2:20   David Davenport -- A Non-Destructive Approach to Reducing Riparian Transpiration
2:40   Daniel Frost -- The Value of Riparian Forests in Today's Society
3:00   Refreshment Break
3:15   Felix Smith -- A Personal Summary
3:30   Panel Discussion and Questions from Invited Guests and the Audience

ACKNOWLEDGMENTS

Riparian Forest Symposium Committee

    Geoffrey A. Wandesforde-Smith: Acting Director - Institute of Ecology

    Anne Sands: Coordinator

    Lyn Schonewise: Secretary

    David Gaines

    J. Greg Howe

    Jack Major

    Warren Roberts

    Robert Robichaux

    Dean Taylor

    Pete Sands: Audio-visual

    Sally and Dean Jue: Refreshments

Publication Staff

    Francisco J. Ayala: Director - Institute of Ecology

    Anne Sands: Editor

    J. Greg Howe: Assistant Editor

    Wanda Greene: Principal Typist

    Dolores DuMont and Cece Odelius: Transcribers

    Cliff Shockney: Typist

    Joyce Ogle: Distribution

Chapter 1

A SHORT REVIEW OF THE STATUS OF RIPARIAN FORESTS IN CALIFORNIA

Felix Smith
Field Supervisor, Division of Ecological Services
United States Fish and Wildlife Service

Riparian vegetation along streambanks, where there is usually fertile soil and an ample water supply, is a most striking feature of California's landscape. These forests appear as a green belt along permanent and intermittent water courses, sloughs, flood plains, overflow channels and oxbows, drainage ditches and lakes.

One can quickly see that the riparian community with its soil, water, and vegetation is a complex ecosystem. Cheatham and Haller (California Fish and Game, 1965), in their "Annotated List of California Habitat Types" have identified four major riparian habitats with 11 subhabitat types. Of the 29 habitat types listed in the "Inventory of Wildlife Resources, California Fish and Wildlife Plan" (Vol. III), riparian habitat provides living conditions for a greater variety of wildlife than any other habitat type found in California. It was estimated in 1963 that riparian vegetation covered about 347,000 acres -- less than one-half of one percent of the total land area of the State.

Factors affecting or adversely impacting riparian vegetation include upstream reservoir construction, levee and channelization projects, and water conservation. The reservoir, levee and channelization activities, along with clearing for agriculture, are common activities that have occurred throughout the State. Removal of vegetation is a common practice in the Colorado River area. Let's look at a riparian area from a local viewpoint. In An Island Called California, Elna Bakker (1971) states that no natural landscape in California has been so altered by man as its bottom lands. It was in the Central Valley that riparian forests were most extensive and were called gallery forests. Coupled with the extensive grasslands and rivers, large and small, a unique setting was created. It is now one of the richest agricultural areas in the world, blessed with good climate, rich soil, and until the last couple of years, ample water supplies.

The Sacramento River from Red Bluff to its mouth in the Delta is a meandering alluvial stream. The Sacramento Valley extends about 150 miles north-south and spreads about 45 miles east-west at its widest point, averaging about 30 miles wide. The area of the Sacramento Valley, so defined, is about 5,000 square miles; the area of the entire Sacramento River drainage basin is 26,150 square miles.

The Sacramento Valley is bounded by the Coast Ranges on the west, the Klamath Mountains on the north, and the southern Cascade Range and northern Sierra Nevadas on the east. The southern margin is extremely low terrain, cut by numerous branching channels of the Delta. This whole low-lying and level area is formed by the combined Delta of the Sacramento and San Joaquin Rivers. Lands of the Sacramento Valley, excluding the Sutter Buttes, are essentially flat, almost featureless, and were formed by the long-continued accumulation of sediments in a great structural trough lying between the Coast Ranges and the Cascades-Sierra Nevada. Large and small streams break up the landscape. Each had its green belt of riparian vegetation that stretched from the base of the foothills to the big river and adjacent wetlands.

Vegetation will grow on any portion of a streambed and its banks if the soil or other substrate is exposed long enough during the growing season. The fertile loam soils of the Sacramento River riparian land coupled with favorable ground water conditions and a long growing season provide near optimum conditions for the establishment of the extensive riparian forests.

The riparian woodlands occurred on the natural levees formed by the Sacramento, Lower Feather, American, and other aggrading streams. These levees rose from 5 to 20 feet above the streambed, and ranged in width from 1 to 10 miles. Based on historical accounts, it has been estimated that there were about 775,000 acres of riparian woodlands in 1848 - 1850. Diaries and field notes written in the early 1800's describe the extent of the forests. They also describe the lush jungles of oak, sycamore, ash, willow, walnut, alder, poplar, and wild grape which comprised almost impenetrable walls of vegetation on both

1

sides of all the major valley rivers and their tributaries. Notes were made of giant sycamores 75 to 100 feet tall and of oaks 27 feet in circumference. By the late 1800's, however, vast tracts of riparian forests had already been cut by settlers for fuel and building materials. In addition, many thousands of acres were cleared to free the fertile alluvial soil for agricultural use. By 1952, only about 20,000 acres of riparian forests remained. Today's estimate of 12,000 acres is probably generous.

Prior to 1960, few people showed any concern for the demise of California's Riparian Forest communities. In addition, very little botanical data had been collected. During the early 60's the first major work at removing the riparian forest remnants in an effort to protect levees occurred in the Delta. The removal of this riparian vegetation from along the lower Sacramento River was viewed with great concern by the public. Statements, both written and oral, voiced strong opposition to the methods of levee maintenance and stated that better methods should be employed so as not to destroy the esthetic beauty and wildlife habitat of the Delta waterways. Most of the same concerns exist today. However, today dedicated and enthusiastic botanists, ornithologists, mammalogists, entomologists and other field scientists are compiling species lists, recording observations, and beginning to publicize their findings. People now realize that public awareness must be coupled with political pressure.

The previously expressed concerns demonstrate a clear need for a higher order of planning and evaluation before additional irreparable alterations to this river system occur. Although no governmental body, agency, interest, or person would deliberately set out to destroy the Sacramento and other California rivers, adjacent lands and natural resources, all too often there has been insufficient concern about the singular or cumulative effects of work accomplished by one agency or interest on the resources under the jurisdiction or responsibility of another, or how such work affects the entire riverine ecosystem and the public interest.

The realization of a Sacramento River environmental/open space corridor is a valid and long-term planning objective. Implementation is the difficult part. However, it can be done. It will require the formulation of a multigovernmental agency and concerned citizen group to see that modifications and developments are accomplished without further deterioration of the existing resources and that efforts are undertaken to enhance these same resources in the public interest while at the same time protecting the integrity of the levees and communities of the Sacramento Valley. Can one imagine a Sacramento River Parkway from the Redding area to Collinsville patterned after the American River Parkway? What a valuable recreational resource it would be to the public and especially for future generations!

Literature Cited

Bakker, Elna S. An Island called California. University of California Press. 1971.

California Fish and Game Plan. California Department of Fish and Game. 1965.

Chapter 2

A SURVEY OF RIPARIAN FOREST FLORA AND FAUNA IN CALIFORNIA

Warren G. Roberts, J. Greg Howe and Jack Major
Departments of Botany and Wildlife and Fisheries Biology
University of California at Davis

California is a desert in the warm half of the year in cismontane California and all the year in transmontane. Even the mountains have precipitation of only desert dimensions in summer. The all-year desert occupies all the southern San Joaquin Valley (Major, 1977), and, even as far north as Williams, the Central Valley receives only 390 mm/yr of precipitation with need for water greater than the upland supply during eight months. Riparian habitats are the only mesic breaks in these very extensive desert tracts. Their extraordinary natural productivity is due to optimum conditions of sunlight, water, and nutrients. As one might expect from the presence of abundant, dense, rapidly transpiring vegetation, the microclimate in riparian forests is similar to that of a greenhouse. Humidity and temperatures are high, the air is still, and evapotranspiration can be rapid. Heller (1969) showed the quantitative dependence of this unique microclimate on the amount of water flowing in the river.

Riparian habitats are azonal. They are not only determined by the climatic zone in which they happen to occur, the kinds of rocks from which their soils have formed, or the available kinds of plants and animals, but also by features peculiar to them. For example, they have a high water table, and nutrients are supplied in abundance both in ionic form and as unweathered rock particles by annual flooding. Their soils are often relatively coarse textured, well drained, and well aerated, at least at the surface horizons. Also, these soils are young, without hardpans, and have periodically open niches as the result of flooding, bank erosion, and deposition of mixed gravels and silts. As well, these alluvial deposits are often naturally fertilized by organic debris from the forests. If clays occur, they are in thin lenses deposited in backwaters, and they usually hold up a water table only temporarily.

The high biological productivity of California riparian habitats can be seen in the following examples. The coast redwoods develop best on riparian floodplain sites producing the largest biomass figure known: 452,500 grams per square meter (Franklin & Dyrness 1973). The montane forests are most magnificent in the valleys. In the foothills, shade from streamside trees contrasts with sun-soaked chaparral slopes. The grandest hardwood forests in California undoubtedly once bordered the Sacramento River. The Fremont cottonwoods of the drier interior Coast Ranges are bright green ribbons in a darker, evergreen, or tawny landscape. In the Great Basin the same tree contrasts even more with sagebrush. In the hot deserts trees may still occur in riparian habitats amidst a landscape of low, widely-spaced, xerophytic shrubs. The Colorado River Delta in the most extreme desert area in North America must have once had an enormous productivity, judging by the description by Aldo Leopold (L. Leopold 1953). In the same desert area dry washes are marked by tall shrubs of very different species from those in the zonal desert landscape.

Ornduff has aptly summarized (1974:99): "Since the climatic regime over much of California is an arid one, the local occurrence of permanent standing or running water has a striking influence on the vegetation. The many large streams and rivers that flow out of the California mountains are generally lined with deciduous trees, shrubs, and herbs that are restricted to the banks of these water courses." These complex communities can also include vines and tall herbaceous undergrowth, and, in the milder and moister parts of California such as the North Coast, the riverine forests include evergreen as well as deciduous species. Some of California's most rare and endangered plant species are limited to the riparian habitat. An example is the beautiful wild hibiscus, Hibiscus californicus, an herbaceous perennial shrub with 15 - 20 cm diameter white and dark red flowers found only on the few undisturbed slough banks which remain in the Central Valley.

Much of what we understand about riparian woodlands and their ecology in California must be interpreted from work done on the ecology of riparian ecosystems in other parts of the world, because relatively few studies have been completed in California. We can, however, provide a general description of California's riparian forests and their plant and animal species based on literature searches and personal communications. Our survey was

limited to depositional (alluviating) streams below 4,000 ft (1,200 m) elevation with perennial surface water (Tables 1, 2, 3 and 4 prepared by Warren G. Roberts and J. Greg Howe).
We did not include the very narrow riparian woodland bordering the scouring, or non-depo-
sitional, streams in the mountains nor the depositional stretches of streams above 1200 m
such as Vidette Meadows on the Kings River, Tuolumne Meadows, and the upper Kern River
meadows. The strictly herbaceous plants of riverine forests were also excluded.

Woody plants of the riparian woodlands of the Central Valley, North Coast, South Coast,
Palm Oases, Deserts, and Northeastern Valleys are listed in Table 1 and illustrated in
Figure 1. The first four types are typically Californian, while the remaining three are
more closely related to forests of adjacent states and Mexico. It should be noted that
some of the plants in the riparian forest originated from cooler, moister ecological zones
which occasionally overlapped the riverine habitat. Examples of these are Quercus agrifolia
in the Sacramento-San Joaquin Delta, and Pinus ponderosa along the Owens River. Adventive
species (plants which have established in the riparian woodland as the result of human
activities) include the following exotic genera: Ailanthus, Robinia, Tamarix, and Arundo.

Although knowledge of riparian woodland invertebrate and vertebrate fauna is limited,
we included partial lists of mammals, reptiles, amphibians, and butterflies known to inhabit
the remaining stands of this extraordinary habitat. An illustration of the urgent need for
more extensive field work is Art Shapiro's discovery that a beetle, Desmocerus californicus
ssp. dimorphos, is known only from the riparian forests of Sacramento, Yolo, and Merced
Counties. It has been nominated for the United States Department of the Interior's
"Threatened List." The beetle requires Sambucus caerulea and S. mexicana as host plants.
The avifauna are excluded from this survey as they are treated in detail by David Gaines
in Chapter Seven. Likewise, the fishes are discussed by Donald Alley in Chapter Eight.
Virtually all of the butterflies known to breed in the Sacramento Valley have been seen
in riparian forests. Those which are most associated with such forests are listed in
Table 4.

The original riparian forests covered several million acres; today they are measured
in the thousands (Figure 2). For example, the Sacramento River had 800,000 acres of riparian vegetation left in 1848 and 12,000 in 1972; five percent of the high terrace habitat
has been lost in the last twenty years. The riparian woodland along the Colorado River has
diminished rapidly in just the last five years. Thus, most of California's riparian ecosystems have been destroyed or degraded. Major man-caused changes have been conversion of
forest to orchard and field crops, logging for wood chips, streambank stabilization, channelization, reduction of water flow by dams and irrigation, and accelerated erosion of river
banks due to dams upstream and channelization in adjacent areas.

Further losses of habitat are attributable to gravel and gold mining, grazing, and
water pollution. Urbanization brings its housing developments, freeways and landfills to
the riverlands. Even recreational developments can have detrimental results: removal
of understory vegetation, construction of docks and boat ramps, and introduction of exotic
plant species all speed the degradation of the riparian ecosystem. In 1958, Wistendahl
optimistically stated: "The increasing recognition of the value of natural areas as part
of park systems makes it possible that certain parts of the flood plain system may ultimate-
ly be allowed to grow back to their original magnificence." We hope these remarks will be
heeded before all the significant riparian species are gone.

Literature Cited

Abrams, L. & R. S. Ferris. 1923-60. Illustrated flora of the Pacific States. 4 vol.
    Stanford Univ. Press, Stanford.

Baker, M. S. 1951. A partial list of seed plants of the north coast counties of California.
    Santa Rosa Junior Coll, Santa Rosa.

Bakker, E. S. 1971. An island called California. Univ. Calif. Press, Berkeley.

Berry, Shad. personal communication.

Franklin, J. F. & C. T. Dyrness. 1973. Natural vegetation of Oregon and Washington.
    U. S. Forest Service, Pacific Northwest Forest & Range Exp. Sta. Gen. Techn. Report
    PNW-8, p. 54.

Griffin, J. R. & W. B. Critchfield. 1972. The distribution of forest trees in California. U. S. Forest Service, Pacific SW For. & Range Exp. Sta. Res. Paper PSW-82.

Heller, H. 1969. Lebensbedingungen und Abfolge der Flussauenvegetation in der Schweiz. Mitt. Schweiz. Anst. forstl. Versuchswesen 45(1):3-124.

Holmes, L. C., Nelson, J. W., and Party. 1915. Reconnaissance Soil Survey of the Sacramento Valley, California. U. S. Dept. of Agriculture Publ., (Government Printing Office) Washington D. C.

Hoover, R. F. 1970. The vascular plants of San Luis Obispo County, California. Univ. Calif. Press, Berkeley.

Howell, J. T. 1970. Marin flora, with supplement. Univ. Calif. Press, Berkeley.

Howitt, B. F. & J. T. Howell. 1964. The vascular plants of Monterey County, California. Wasmann J. Biol. 22(1):1-184. 1973, Supplement, Pacific Grove Mus. Natural History Assoc, Pacific Grove, California.

Ingles, L. G. 1965. Mammals of the Pacific States. Stanford Univ. Press, Stanford.

Jepson, W. L. 1893. The riparian botany of the lower Sacramento. Erythea 1:228-246.

Jepson, W. L. 1909-1943. A flora of California. Vols. 1-3. Assoc. Students Store, Univ. California, Berkeley.

Jue, Dean. Personal communication.

Leopold, L. B., ed. 1953. Round river. Oxford Univ. Press, N. Y., pp. 10-30.

Lowe, Charles H., Personal communication.

Major, J. 1977. California climate in relation to vegetation. In Terrestrial vegetation of California, ed. M. G. Barbour & J. Major. pp. 11-74. John Wiley, N. Y.

Major, J. 1963. Checklist of vascular plants in Yolo, Sacramento, Solano and Napa Counties, California. Dept. Botany, Univ. California, Davis.

Mason, H. L. 1957. A flora of the marshes of California. Univ. Calif. Press, Berkeley.

Munz, P. A. 1974. A flora of southern California. Univ. Calif. Press, Berkeley.

Munz, P. A. 1968. Supplement to a California flora. Univ. Calif. Press, Berkeley.

Munz, P. A. & D. D. Keck. 1959. A California flora. Univ. Calif. Press, Berkeley.

Nelson, J. W., J. E. Guerney, L. C. Holmes, and E. C. Eckman. 1918. Reconnaissance Soil Survey of the Lower San Joaquin Valley, California. U. S. Dept. of Agriculture Publ., Government Printing Office, Washington D. C.

O'Brien, R. J., L. K. Puckett, & T. B. Stone. 1976. Birds in riparian habitat of the upper Sacramento River. Memorandum Report, Calif. Dept. Fish & Game, Sacramento.

Ornduff, R. 1974. Introduction to California plant life. Univ. Calif. Press, Berkeley.

Parish, S. B. 1930. Vegetation of the Mohave and Colorado deserts of southern California. Ecology 11:481-503.

Peck, M. E. 1961. A manual of the higher plants of Oregon. Oregon State Univ. Press, Corvallis.

Rudd, Robert. Personal communication.

Sacramento County Office of Education. 1975. The outdoor world of the Sacramento region.

Shapiro, A. M. 1974. The butterfly fauna of the Sacramento Valley, California. J. Res. Lepidoptera 13:73.

Stebbins, R. C. 1966. A field guide to western reptiles and amphibians. Houghton Mifflin Co., Boston.

Stone, T. 1976. Observations on furbearers within the riparian habitat of the upper Sacramento River, California. Calif. Fish & Game Memorandum Report.

Sudworth, G. B. 1908. Forest trees of the Pacific slope. U. S. Forest Service Bull. Dover ed., 1967.

Taylor, Dean. Personal communication.

Trapp, Gene. Personal communication.

Twisselmann, E. C. 1967. A flora of Kern County, California. Wasmann J. Biology 25(1/2):1-395.

Vogl, R. T. & L. T. McHargue. 1966. Vegetation of California fan palm oases on the San Andreas Fault. Ecology 47:532.

Wistendahl, W. A. 1958. The flood plain of the Raritan River, New Jersey. Ecol. Monogr. 28:129-153.

Figure 1. Riparian Woodland Divisions of California.

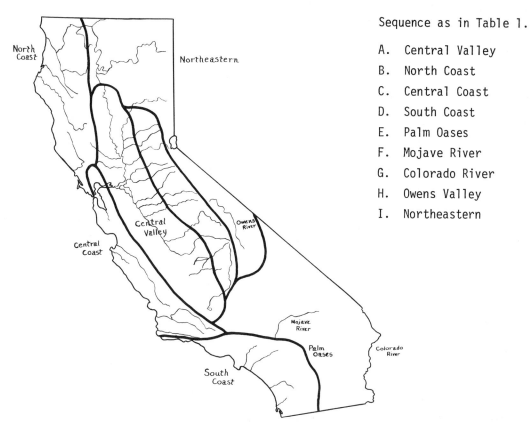

Sequence as in Table 1.

A. Central Valley
B. North Coast
C. Central Coast
D. South Coast
E. Palm Oases
F. Mojave River
G. Colorado River
H. Owens Valley
I. Northeastern

Figure 2. The original (in gray) and existing (in black) Riparian Woodland of the Sacramento and San Joaquin River Drainages (depositional portions only) compiled by J. Greg Howe.

The original extent of riparian woodland was determined with the use of soil maps published before 1920 (Holmes, 1915 and Nelson, 1918). These maps, in particular the reconnaissance soil survey maps of the Sacramento Valley and the lower San Joaquin Valley, outline all soil types and describe their existing and pristine native vegetation. An example of this soil analysis is the Columbia series. The two Columbia soils, Columbia fine sandy loam and Columbia silt loams, are the principle soil groups lining the Sacramento and Feather Rivers and make up most of the gray region on the map.

A description of the native vegetation on the Columbia fine sandy loam describes it as "originally covered with a heavy forest growth consisting mainly of sycamore, cottonwood, willow and oak, with a thick undergrowth. A great deal of the surface has been cleared and cultivated, but much of the lowest. . . areas yet remain a thick jungle." (Holmes, 1915 p. 108 Ref. 1) And also, referring to the Columbia silt loams; "originally the soils of this group were heavily timbered. In the better drained locations, slightly removed from the stream channels, the valley oak predominated with a vigorous undergrowth. The lower-lying areas, or those most subject to overflow, supported a tangled growth approaching a tropical jungle in density, consisting mostly of cottonwood, sycamore, willow and wild grape." (Holmes, 1915, p. 111) The same descriptions occur in the Lower San Joaquin Valley soil survey; however, they are not as detailed and inviting as those of the Sacramento soil groups.

The present distribution of riparian woodland on the major rivers was developed from aerial photographs, most of which were taken after 1970. The San Joaquin and Sacramento river photographs were taken in 1975. All photographs (with scales ranging from 1:6,000 - 1:24,000): the California Department of Water Resources and the U.S. Army Corps of Engineers. Some of the smaller rivers, such as Stony and Honcut creek, were analyzed for woodland with the use of photo-revised (1969 or later) United States Geological Survey maps.

Figure 2.

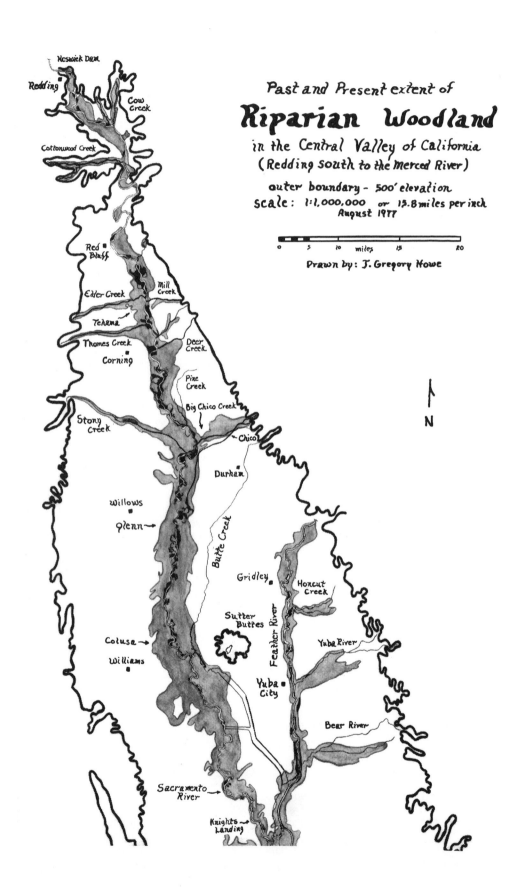

Past and Present extent of
Riparian Woodland
in the Central Valley of California
(Redding south to the Merced River)
outer boundary - 500' elevation
scale: 1:1,000,000 or 15.8 miles per inch
August 1977

Drawn by: J. Gregory Howe

N

Figure 2.

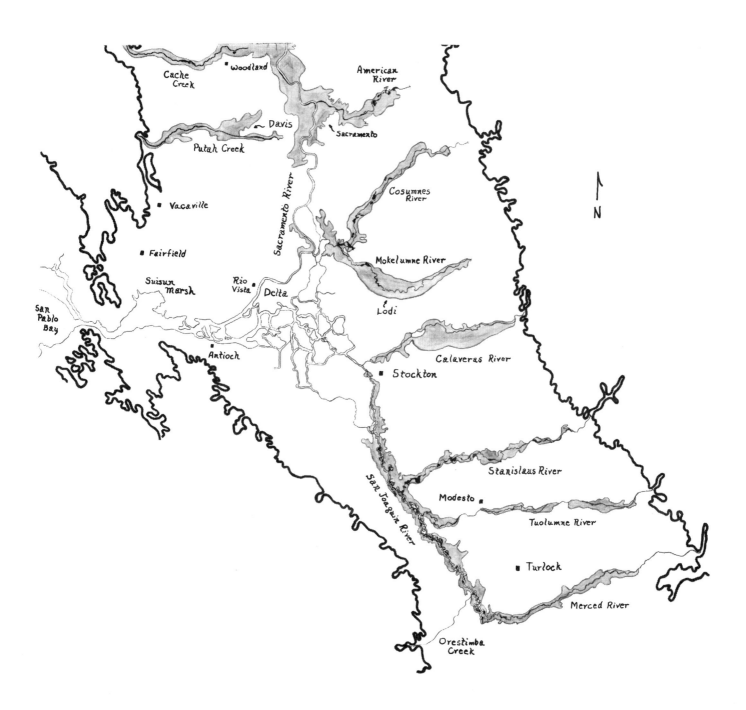

Table 1. Woody Plants of the Riparian Forests of California.

A. Central Valley Riparian Forest.

  1. TREES

    a. Common:

      Acer negundo subsp. *californicum*. box elder.
      Platanus racemosa. California sycamore.
      Populus fremontii. cottonwood
      Quercus lobata. valley oak, water oak.
      Salix goodingii var. *variabilis*. willow.
      Salix laevigata. red willow.
      Salix lasiandra. black willow.

    b. Uncommon:

      Aesculus californica. California buckeye.
      Ailanthus altissima. tree-of-heaven. (exotic)
      Alnus rhombifolia. sierra alder.
      Ficus carica. fig. (exotic)
      Fraxinus latifolia. ash.
      Juglans hindsii. native black walnut.
      Morus alba. white mulberry. (Kern R.)
      Pistacia chinensis. Chinese pistache. (Chico area) (exotic)
      Quercus agrifolia. coast live oak. (Delta area)
      Quercus wislizenii. interior live oak.
      Robinia pseudo-acacia. black locust. (exotic)
      Salix goodingii var. *goodingii*. willow. (s. from Yolo Co.)

  2. SHRUBS

    a. Common:

      Artemisia douglasiana. mugwort.
      Arundo donax. giant reed. (exotic)
      Baccharis viminea. mule fat, false willow.
      Cephalanthus occidentalis. button-willow.
      Phoradendron tomentosum subsp. *macrophyllum*. big mistletoe. (parasite)
      Phragmites communis var. *berlandieri*. common reed.
      Rosa californica. wild rose.
      Rubus discolor (syn. *R. procerus*). Himalayan blackberry. (exotic)
      Rubus ursinus. wild blackberry.
      Rubus vitifolius. wild blackberry.
      Salix hindsiana. sandbar willow.
      Salix lasiolepis. willow.
      Salix melanopsis. willow.
      Sambucus mexicana. elderberry.
      Symphoricarpos rivularis. snowberry.
      Tamarix parviflora. tamarisk, salt-cedar. (exotic)

    b. Uncommon:

      Atriplex lentiformis. quail-bush. (San Joaquin V.)
      Baccharis douglasii. false-willow.
      Baccharis glutinosa. false-willow. (s. San Joaquin V.)
      Cornus glabrata. brown dogwood.
      Cornus occidentalis. red osier dogwood.
      Heteromeles arbutifolia. toyon.
      Hibiscus californicus. wild hibiscus.
      Lonicera involucrata. twin-berry honeysuckle. (delta)
      Ptelea crenulata. hop tree. (n. Sacramento V.)
      Rubus laciniatus. cut-leaf blackberry. (exotic)

Table 1, continued.

3. VINES

Aristolochia californica. Dutchman's pipe vine. (Sacramento V.)
Clematis lasiantha. wild clematis.
Clematis ligusticifolia. wild clematis.
Lonicera hispidula var. *vacillans.* wild honeysuckle. (uncommon)
Rhus diversiloba. poisonoak.
Smilax californica. greenbrier. (n. Sacramento V.)
Vitis californica. wild grape.

B. North Coast Riparian Forest.

1. TREES

a. Common:

Acer macrophyllum. big-leaf maple.
Alnus rubra. red alder.
Populus trichocarpa. black cottonwood.
Salix laevigata. willow.
Salix lasiandra. willow.
Sequoia sempervirens. redwood
Umbellularia californica. bay, pepperwood.

b. Uncommon:

Abies grandis. lowland fir.
Aesculus californica. California buckeye.
Alnus rhombifolia. white alder. (inland from the fog belt)
Picea sitchensis. Sitka spruce.
Thuja plicata. red cedar.
Tsuga heterophylla. western hemlock.

2. SHRUBS

a. Common:

Artemisia douglasiana. mugwort.
Artemisia suksdorfii. mugwort.
Baccharis douglasii. false willow.
Baccharis viminea. false willow.
Cornus occidentalis. red osier dogwood.
Physocarpus capitatus. ninebark.
Rhododendron occidentale. western azalea.
Rosa californica. wild rose.
Rubus discolor. Himalayan blackberry. (syn. *R. procerus*) (exotic)
Rubus ursinus. wild blackberry.
Rubus vitifolius. wild blackberry.
Salix coulteri. willow.
Salix hindsiana. sandbar willow.
Salix lasiolepis. willow.
Salix melanopsis. willow.
Salix scouleriana. willow.
Sambucus caerulea. elderberry.
Sambucus callicarpa. red elderberry.

b. Uncommon:

Acer circinatum. vine maple.
Calycanthus occidentalis. spicebush.
Cornus glabrata. brown dogwood.
Euonymus occidentalis. native euonymus.
Lonicera involucrata. twinberry honeysuckle.
Mahonia nervosum. dwarf Oregon-grape.
Malus fusca. wild crabapple.
Myrica californica. bayberry, California wax-myrtle.

11

Table 1, continued.

Phragmites communis var. *berlandieri*. common reed.
Ribes bracteosum. stink currant. (n. from Mendocino Co.)
Ribes divaricatum. wild gooseberry.
Ribes laxiflorum. wild currant. (n. from Humboldt Co.)
Ribes menziesii. canyon gooseberry.
Ribes sericeum. wild gooseberry.
Rosa nutkana. wild rose. (n. from Mendocino Co.)
Rubus laciniatus. cut-leaf blackberry. (exotic)
Rubus parviflorus. thimble-berry.
Rubus spectabilis. salmon-berry.
Salix delnortensis. willow. (between 300' and 600' el., Smith R.)
Salix parksiana. willow. (n. from Humboldt Co.)
Salix sitchensis. willow.
Salix tracyi. willow. (n. from Humboldt Co.)
Vaccinium ovatum. huckleberry.

3. VINES

Aristolochia californica. Dutchman's pipe vine. (inland from fog belt)
Clematis ligusticifolia. wild clematis.
Lonicera hispidula var. *vacillans*. wild honeysuckle.
Rhus diversiloba. poisonoak.
Smilax californica. greenbrier. (uncommon)

C. Central Coast Riparian Forest. (north from the Santa Ynez Mountains through the San Francisco Bay area)

1. TREES

   a. Common:

   Platanus racemosa. sycamore.
   Populus fremontii. cottonwood.
   Populus trichocarpa. black cottonwood. (replaces *P. fremontii* on the Carmel R. and is dominant on the Santa Inez R.)
   Salix laevigata. willow.
   Salix lasiandra. willow.

   b. Uncommon:

   Acer macrophyllum. big-leaf maple.
   Acer negundo subsp. *californicum*. box elder. (abundant locally)
   Aesculus californica. California buckeye.
   Alnus rhombifolia. white alder.
   Alnus rubra. red alder.
   Fraxinus latifolia. Oregon ash. (occasional: San Francisco Bay area)
   Juglans hindsii. native black walnut. (native in e. S. F. Bay area counties)
   Prunus cerasifera. cherry plum. (exotic, occasional in e. Marin Co. and s. Sonoma Co.)
   Quercus agrifolia. live oak.
   Quercus lobata. valley oak.
   Sequoia sempervirens. redwood.
   Umbellularia californica. baytree, California laurel.

2. SHRUBS

   a. Common:

   Artemisia douglasii. mugwort.
   Arundo donax. giant reed. (exotic)
   Baccharis douglasii. false willow.
   Baccharis viminea. false willow, mule fat.
   Cornus occidentalis. red osier dogwood.
   Phoradendron tomentosum subsp. *macrophyllum*. big mistletoe. (parasite)
   Phragmites communis var. *berlandieri*. common reed.
   Rosa californica. wild rose.
   Rubus ursinus. wild blackberry.

12

Table 1, continued.

> *Rubus vitifolius.* wild blackberry.
> *Salix coulteri.* willow.
> *Salix hindsiana.* willow.
> *Salix lasiolepis.* willow.
> *Salix lasiolepis* var. *bigelovii.* willow.
> *Salix melanopsis.* willow.
> *Salix scouleriana.* willow.
> *Sambucus mexicana.* elderberry.
> *Symphoricarpos rivularis.* snowberry.

    b. Uncommon:

> *Atriplex lentiformis* subsp. *breweri.* quailbush.
> *Cornus glabrata.* brown dogwood.
> *Euonymus occidentalis.* native euonymus.
> *Forestiera neomexicana.* desert olive.
> *Lonicera involucrata.* twin berry honeysuckle.
> *Myrica californica.* wax myrtle.
> *Physocarpus capitatus.* nine bark.
> *Rhododendron occidentale.* western azalea.
> *Ribes aureum* var. *gracillimum.* golden currant.
> *Ribes divaricatum.* wild gooseberry.
> *Ribes menziesii.* canyon gooseberry.
> *Ribes sericeum.* wild gooseberry.
> *Rubus discolor* (syn. *R. procerus*). Himalayan blackberry. (exotic)
> *Rubus laciniatus.* cut-leaf blackberry. (exotic)
> *Rubus parviflorus.* thimbleberry.
> *Sambucus callicarpa.* red elderberry.

  3. VINES

> *Aristolochia californica.* Dutchman's pipe vine. (uncommon).
> *Clematis lasiantha.* wild clematis.
> *Clematis ligusticifolia.* wild clematis.
> *Hedera helix.* English ivy. (exotic—occasional escape)
> *Lonicera hispidula* var. *vacillans.* wild honeysuckle.
> *Rhus diversiloba.* poisonoak.

D. South Coast Riparian Forest. (south from the Santa Ynez Mountains)

  1. TREES

    a. Common:

> *Alnus rhombifolia.* white alder.
> *Platanus racemosa.* native sycamore, aliso. (dominant)
> *Populus fremontii.* cottonwood, álamo.
> *Quercus agrifolia.* coast live oak, encina.
> *Salix laevigata.* willow, sauce.
> *Salix lasiandra.* willow, sauce.

    b. Uncommon

> *Acer macrophyllum.* big-leaf maple.
> *Acer negundo* subsp. *californicum.* box elder.
> *Ailanthus altissima.* tree-of-heaven. (exotic)
> *Juglans californica.* native black walnut, nogal.
> *Populus trichocarpa.* black cottonwood, álamo.
> *Umbellularia californica.* California laurel, bay tree, laurel.

  2. SHRUBS

    a. Common

> *Artemisia douglasiana.* mugwort.
> *Arundo donax.* giant reed, carrizo. (exotic)

Table 1, continued.

Baccharis emoryi. baccharis.
Baccharis glutinosa. water-wally, seep willow.
Baccharis pilularis subsp. *consanguineus.* coyote bush.
Baccharis sarothroides. broom baccharis.
Baccharis viminea. mule fat.
Cornus occidentalis. red osier dogwood.
Phoradendron tomentosum subsp. *macrophyllum.* big mistletoe. (parasite)
Phragmites communis var. *berlandieri.* common reed, carrizo.
Rosa californica. wild rose, rosa.
Rubus ursinus. wild blackberry.
Salix hindsiana. willow, sauce.
Salix lasiolepis. willow, sauce. (the most common willow in s. Calif.)
Sambucus mexicana. elderberry, saúco.

b. Uncommon:

Atriplex lentiformis subsp. *breweri.* quail bush. (n. from Orange Co.)
Baccharis douglasii. mule fat, false willow.
Berberis nevinii (syn. Mahonia n.). San Fernando barberry.
Cornus glabrata. brown dogwood.
Cornus stolonifera. red osier dogwood.
Forestiera neomexicana. desert olive.
Myrica californica. wax myrtle.
Ribes aureum var. *gracillimum.* golden currant.
Salix goodingii var. *variabilis.* willow, sauce.
Tamarix africana. tamarisk. (mouth of Ventura R.) (exotic)
Tamarix parviflora. tamarisk. (exotic)

3. VINES

Clematis ligusticifolia. wild clematis.
Lonicera hispidula var. *vacillans.* wild honeysuckle. (n. from W. Riverside Co., uncommon)
Rhus diversiloba. poisonoak.
Vitis girdiana. wild grape, uva cimarrona.

E. Palm Oases.

1. TREES

Cercidium floridum. palo verde.
Platanus racemosa. sycamore.
Populus fremontii. cottonwood.
Washingtonia filifera. California fan palm.

2. SHRUBS

Arundo donax. giant reed. (exotic)
Atriplex lentiformis. quail bush.
Baccharis glutinosa. water-wally, seep-willow.
Baccharis sarothroides. broom baccharis.
Chilopsis linearis. desert willow.
Nerium oleander. oleander. (uncommon) (exotic)
Phragmites communis var. *berlandieri.* common reed.
Pluchea sericea. arrowweed.
Prosopis glandulosa. mesquite.
Prosopis pubescens. screwpod.
Salix exigua. narrow-leaf willow.
Salix goodingii var. *goodingii.* willow.

3. VINES

Vitis girdiana. wild grape.

14

Table 1, continued.

F. Mojave River Riparian Forest.

   1. TREES

   Alnus rhombifolia. alder. (to Victorville)
   Fraxinus velutina. ash. (to Victorville)
   Platanus racemosa. sycamore. (to hesperia)
   Populus fremontii. cottonwood. (to Alton)

   2. SHRUBS

   Atriplex lentiformis. quailbush.
   Baccharis glutinosa. water-wally, seep-willow.
   Phragmites communis var. *berlandieri.* reed.
   Salix exigua. narrow-leaf willow.
   Salix goodingii var. *goodingii.* willow. (to Alton)

   3. VINES

   ? Vitis girdiana. wild grape.

G. Colorado River Riparian Forest, woody plant species.

   1. TREES

   ? Parkinsonia aculeata. Jerusalem thorn. (exotic)
   Populus fremontii var. *macdougallii.* cottonwood, álamo.
   Tamarix aphylla. salt-cedar. (common only at Needles) (exotic)

   2. SHRUBS

   Arundo donax. giant reed. (exotic)
   Baccharis glutinosa. water-wally.
   Baccharis sarothroides. broom baccharis.
   Phragmites communis var. *berlandieri.* common reed.
   Pluchea sericea. arrowweed.
   Salix exigua. narrow-leaf willow.
   Salix goodingii var. *goodingii.* willow, sauce.
   Sesbania macrocarpa. Colorado River hemp. (woody annual)
   Tamarix chinensis. tamarisk. (exotic)
   Tamarix ramosissima. tamarisk. (exotic)

   3. MESQUITE ASSOCIATION. (restricted to a narrow belt along the outer edge of the riparian area and mostly above the high water mark. It has been largely destroyed for fuel)

   Atriplex lentiformis. quailbush.
   Cercidium floridum. palo verde.
   Chilopsis linearis. desert willow.
   Lycium spp. box thorn.
   Prosopis glandulosa. mesquite.
   Prosopis pubescens. screwpod, tornillo.
   Suaeda torreyana. seep weed.

H. Owens Valley Riparian Forest. woody plant species. The Owens River is an active depositional stream for about 40 miles downstream from Crowley Lake, continually producing new channels. It is said to be bordered for much of its length with good riparian forest including some very large *Populus fremontii.*

   1. TREES

   Celtis reticulata. hackberry. (uncommon)
   Pinus ponderosa. yellow pine. (uncommon)
   Populus fremontii. cottonwood. (dominant)

15

Table 1, continued.

Populus trichocarpa. black cottonwood.
Salix laevigata. willow. (dominant)

2. SHRUBS

Baccharis glutinosa. water-wally
Betula occidentalis. western birch. (above 4000' el.) (uncommon)
Forestiera neomexicana. desert olive. (uncommon)
Phragmites communis var. *berlandieri.* reed.
Rosa woodsii. wild rose.
Salix exigua. narrow-leaf willow.
Salix lasiolepis var. *bracelinae.* willow.
Salix ligulifolia. willow.
Tamarix sp. tamarisk. (uncommon) (exotic)

I. Northeastern California Riparian Forests, woody plant species. (within the broad valleys in the Cascade and Siskiyou ranges and the Great Basin)

1. TREES

Alnus rhombifolia. alder.
Fraxinus latifolia. ash.
Populus trichocarpa. cottonwood.
Quercus garryana. Oregon oak. (Siskiyou Range)
Salix laevigata. willow.

2. SHRUBS

Cornus stolonifera. red osier dogwood.
Crataegus douglasii. western hawthorn.
Phragmites communis var. *berlandieri.* reed.
Rosa californica. wild rose.
Salix exigua. narrow-leaf willow.
Salix lasiolepis var. *bracelinae.* willow.
Salix lutea var. *watsonii.* willow.

References

Lowe, Charles H. (personal communication)
Taylor, Dean W. (personal communication)
see also Literature Cited

Table 2.   Mammals of the Sacramento Valley Riparian Forests. (Ingles, 1965; Stone, 1976; S. Berry, R. Rudd and G. Trapp, personal communications).

Didelphis marsipialis - common opossum

Scapanus orarius - coast mole

Sorex ornatus - ornate shrew

Myotis yumanensis - Yuma myotis

Myotis californicus - California myotis

Eptesicus fuscus - big brown bat

Lasiurus borealis - red bat

Lasiurus cinereus - hoary bat

Plecotus townsendii - lump-nosed bat

Antrozous pallidus - pallid bat

Tadarida brasiliensis - Brazilian free-tailed bat

Lepus californicus - black-tailed hare

Sylvilagus audubonii - Audubon cottontail

Otospermophilus beecheyi - Beechey ground squirrel

Sciurus griseus - western gray squirrel

Thomonys bottae - Botta pocket gopher

Caster canadensis - beaver

Reithrodontomys megalotis - western harvest mouse

Microtus californicus - California meadow mouse

Ondatra zibethica - muskrat

Rattus norvegicus - Norway rat

Rattus rattus - black rat

Mus musculus - house mouse

Canis latrans - coyote

Vulpus fulva - red fox

Urocyon cinerargenteus - gray fox

Bassariscus astutus - ringtail

Procyon lotor - raccoon

Mustela frenata - long-tailed weasel

Mustela vison - mink

Taxidea taxus - badger

Spilogale putorius - spotted skunk

Mephitis mephitis - striped skunk

Lutra canadensis - river otter

Lynx rufus - bobcat

Sus scrofa - wild boar

Odocoileus hemionus - mule deer

Felis cattus - feral house cat

Erethizon dorsatum - porcupine

Table 3. Reptiles and Amphibians of the Sacramento Valley Riparian Forests.
(Stebbins, 1966; and D. Jue, personal communication).

Ambystoma tigrinum - tigar salamander	rare
Scaphiopus hammondi - western spadefoot	rare
Bufo boreas - western toad	abundant
Hyla regilla - Pacific tree frog	abundant
Rana catesberana - bullfrog	abundant
Rana aurora - red legged frog	?
Clemmys marmorata - western pond turtle	frequent
Sceloporus occidentalis - western fence lizard	abundant
Uta stansburiana - side-blotched lizard	rare
Phrynosoma coronatum - coast horned lizard	rare
Eumecus skiltonianus - western skink	frequent
Eumecus gilbrti - Gilbert's skink	frequent
Cnemidophorus tigris - western whiptail	frequent
Gerrhonotus multicarinatus - southern alligator lizard	common-abundant
Coluber constrictor - racer	common-abundant
Pituophis melanoleucus - gopher snake	common-abundant
Lampropeltis getulus - common kingsnake	frequent
Thamnophis couchi ssp. *gigas* - giant garter snake	endangered-very rare
Crotalus vividis - western rattlensnake	frequent
Thamnophis sirtalis - common garter snake	common

Table 4. Butterflies of the Sacramento Valley Riparian Forests. Data from
Shapiro (1974).

Butterfly			Host Plant
Phyciodes campestris Field Crescent	endemic native	uc	*Aster chilensis*
Polygonia satyrus Satyr Anglewing	native	uc	*Urtica holosericea*
Nymphalis antiops Mourning Cloak	native	fc-c	*Salix* sp.
Limenitis lorquini Lorquins Admiral	endemic native	c	*Salix* sp.
Atlides halesus Great Blue Hairstreak	native	c	*Phoradendron tomentosum* var. *macrophyllum*
Satyrium californica California Hairstreak	endemic native	uc	*Quercus lobata* may be endangered
Satyrium sylvinus Willow Hairstreak	native	c-a	*Salix hindsiana*
Everes comyntes Tailed Blue	endemic native	c	Many species, the most closely associ- ated with riparian woodland being *Lathyrus jepsonii* ssp. *californicus*
Glaucopsyche lygdanus behrii Behr's Silvery Blue	native	c	
Battus philenor Pipevine Swallowtail	native	fc	*Aristolochia californica*
Papilio zelicson Anise Swallowtail	native?	c	*Foeniculum vulgare*
Papilio rutulus Western Tigar Swallowtail	native	c	*Fraxinus, Prunus, Populus* or *Platanus*
Papilio multicaudatus Two-tailed Swallowtail	native	fc	Unknown; likely to be the same as *Papilio rutulus*
Epargyraus clarus Silver Spotted Skipper	native	uc	*Robinia pseudoacacia*
Erynnis persius Persius Duskywing	native	fc	*Lotus purshianus*
Ochlodes sylvanoides Woodland Skipper	native	a	Gramineae
Poanes melane Umber Skipper	native	fc-c	(Gramineae)

Key: a = abundant
 c = common
 fc = frequent-common
 uc = uncommon

Chapter 3

GEOLOGIC HISTORY OF THE RIPARIAN FORESTS OF CALIFORNIA

Robert Robichaux
Department of Botany
University of California, Davis

The modern plant communities of California are the products of evolutionary processes acting over long periods of geologic time. As individual species have evolved and migrated in response to the changing environments, the corresponding plant communities have changed in composition and distribution. The principal evidence for these changes derives from the study of the numerous fossil floras now known from the western United States. Analysis of these floras involves the determination of their species composition and modern geologic occurrence and the reconstruction of their former topographic, climatic, and vegetational settings. Regional comparison of sequences of these floras may then provide evidence for the rates and directions of change in the individual lineages and communities (Axelrod, 1967b).

The history of the vegetation of California has recently been outlined by Axelrod (1977). In separate essays, he has also considered the evolution of desert vegetation, the Sierran Sequoiadendron forest, the closed-cone pine forest, the Santa Lucia fir forest, the montane coniferous forests, and the various sclerophyllous communities of California (Axelrod, 1950, 1959, 1967, 1972, 1975, 1976a,b, 1977). Three general principles emerge from his discussions. First, the modern plant communities of California are composed of taxa of diverse floristic sources. Two principal floristic elements are a southern "Madro-Tertiary" element that includes species in such genera as Arbutus, Arctostaphylos, Ceanothus, Cercocarpus, Cupressus, Quercus, and Umbellularia, and a northern "Arcto-Tertiary" element that includes species in such genera as Acer, Alnus, Castanopsis, Fraxinus, Picea, Quercus, and Sequoia. Second, the modern communities are relatively impoverished representatives of richer, more generalized ancestral communities that included taxa related to species now found only in the southwestern United States and northern Mexico, the eastern United States, or eastern Asia. These "exotic" taxa were gradually eliminated from this region during the late Tertiary in response to a general trend to cooler and drier climate, to a shift in the seasonal distribution of precipitation, and to progressively decreasing equability (Axelrod, 1968). Third, some of the species that are associated in these modern communities have apparently been associated, as ancestral forms in fossil communities, throughout most of California's late Tertiary and Quarternary history, covering a time span of nearly 20 million years. The history of California's riparian forest, as summarized in this assay, illustrates these three general principles.

For the purposes of this discussion, the riparian community along the lower Sacramento River will be considered as the modern starting point. A list of the dominant woody species in this community and their fossil counterparts is provided in Table 1. To facilitate the discussion, we will consider first the present and past distributions of some of these taxa, making some inferences as to their possible geographical sources. We will then consider the history of the riparian community in California during the last 20 million years as recorded in some selected fossil floras from different areas of the state and in nearby regions. The relationship of this information to the future of

Table 1. Some common species in the modern riparian forest of the Sacramento River and their counterparts in the late Tertiary fossil record of the western United States.

Modern Species[1]	Fossil Species[2]
Acer negundo	A. minor
Alnus rhombifolia	A. hollandiana, A. merriami
Cornus californica	C. ovalis
Fraxinus latifolia	F. coulteri, F. caudata
Juglans hindsii	J. pseudomorpha
Platanus racemosa	P. paucidentata
Populus fremontii	P. prefremontii
Quercus lobata	Q. prelobata, Q. moragensis
Salix lasiandra	S. hesperia
Salix lasiolepis	S. wildcatensis
Salix laevigata	S. laevigatoides
Salix goodingii	S. truckeana
Salix hindsiana	S. endenensis
Toxicodendron diversiloba	T. franciscana

[1]Nomenclature follows Munz (1959).

[2]The leaf and seed impressions of the fossil species are generally indistinguishable from those of the modern counterparts. A different name is assigned to the fossil taxon to avoid the inherent difficulties of equating the modern and fossil species (see Axelrod, 1967a).

the riparian forest in California will be our final consideration.

Distributions of the Species

One of the most important factors facilitating a species entry into the fossil record is a proximity to a site of sedimentation. As many fossil deposits accumulate along stream and lake borders, riparian taxa are generally well-represented in the record. Table 1 shows that many of the dominant species in the modern riparian community of the Sacramento River have counterparts in the fossil record of the western United States. The present and past distributions of eight of these species are particularly informative in terms of understanding the floristic sources of the modern forest. These include Acer negundo, Alnus rhombifolia, Fraxinus latifolia, Platanus racemosa, Populus fremontii, Quercus lobata, Salix lasiandra, and Salix lasiolepis (Figs. 1-5).

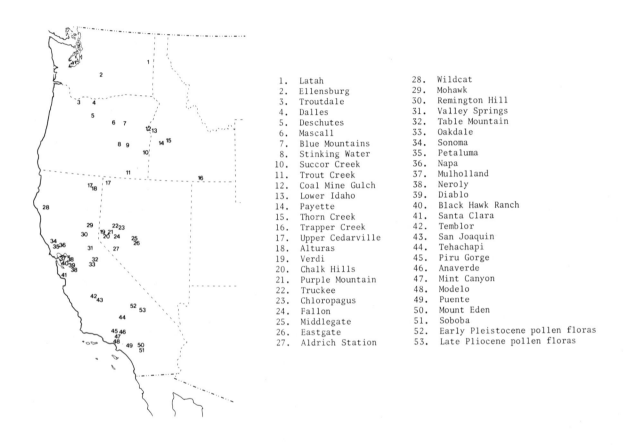

1.	Latah	28.	Wildcat
2.	Ellensburg	29.	Mohawk
3.	Troutdale	30.	Remington Hill
4.	Dalles	31.	Valley Springs
5.	Deschutes	32.	Table Mountain
6.	Mascall	33.	Oakdale
7.	Blue Mountains	34.	Sonoma
8.	Stinking Water	35.	Petaluma
10.	Succor Creek	36.	Napa
11.	Trout Creek	37.	Mulholland
12.	Coal Mine Gulch	38.	Neroly
13.	Lower Idaho	39.	Diablo
14.	Payette	40.	Black Hawk Ranch
15.	Thorn Creek	41.	Santa Clara
16.	Trapper Creek	42.	Temblor
17.	Upper Cedarville	43.	San Joaquin
18.	Alturas	44.	Tehachapi
19.	Verdi	45.	Piru Gorge
20.	Chalk Hills	46.	Anaverde
21.	Purple Mountain	47.	Mint Canyon
22.	Truckee	48.	Modelo
23.	Chloropagus	49.	Puente
24.	Fallon	50.	Mount Eden
25.	Middlegate	51.	Soboba
26.	Eastgate	52.	Early Pleistocene pollen floras
27.	Aldrich Station	53.	Late Pliocene pollen floras

Figure 1. Geographical position of fossil floras from the western United States that contain ancestors of California's lowland riparian taxa.

Acer negundo, the box elder, is composed of several races distributed throughout much of the United States and south-central Canada (Little, 1971). The subspecies californicum is restricted to California (Griffin & Critchfield, 1972) and occurs primarily in the Coast Ranges and along the Sacramento and San Joaquin Rivers of the Central Valley (Fig. 2). The fossil counterpart, Acer minor, was a common tree on the Columbia Plateau and in border areas in the Miocene (Chaney & Axelrod, 1959) when its distribution reached southward to the vicinity of the Middlegate flora of Nevada (Axelrod, 1956) and the Remington Hill and Valley Springs floras of California (Condit, 1944a; Axelrod, 1944e). It survived into the Pliocene in Oregon, as evidenced by the Deschutes and Dalles floras (Chaney, 1938, 1944a). Although it is still unrecorded from the California Pliocene (Chaney & Axelrod, 1959), samaras and leaflets of the modern A. negundo appear in the early Pleistocene Soboba flora of southern California (Axelrod, 1966).

The white alder, Alnus rhombifolia, ranges from California northward along the eastern slopes of the Cascade Ranges of Oregon and Washington to southern British Columbia and

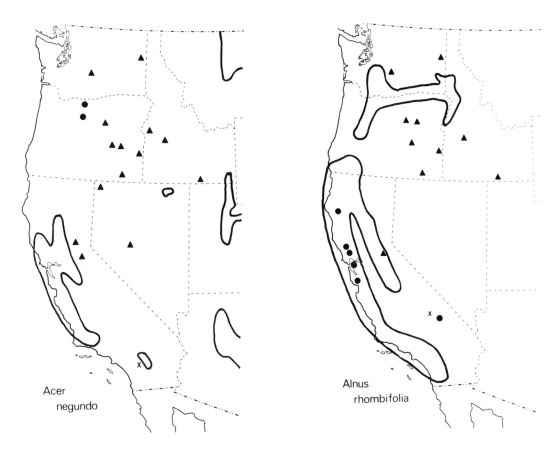

Figure 2. Past and present geographical distributions of Acer negundo and Alnus rhombifolia. The fossil localities correspond to those of Fig. 1. The symbols are: Miocene floras (triangles), Pliocene floras (circles), Pleistocene floras (crosses), and present distributions (solid lines).

northern Idaho (Hitchcock et al., 1964; Little, 1976; Sudworth, 1908). In California, this species is distributed from the Klamath region southward through the Coast ranges and along the western slopes of the Sierra Nevada to the Cuyamaca Mountains of southern California (Fig. 2) (Griffin & Critchfield, 1972). Two fossil species have been referred to this species. A. hollandiana was a regular component of the Miocene forests of the Columbia Plateau and adjacent regions (Chaney & Axelrod, 1959), while A. merriami appears in several of the Pliocene floras of central California (Dorf, 1930; Axelrod, 1944b, d). These two fossil species appear to be closely related and may not be specifically distinct (Chaney & Axelrod, 1959). Pollen similar to the modern A. rhombifolia is recorded from several late Pliocene and early Pleistocene pollen floras from the southern Sierra Nevada (Axelrod & Ting, 1960, 1961).

The modern distribution of Fraxinus latifolia, the Oregon ash, resembles that of the white alder in extending from California northward to southern British Columbia. It differs from the latter in its coastal rather than interior position, with the Cascade ranges forming the eastern and western boundary of the distribution of the ash and alder, respectively, in Oregon and Washington (Little, 1971; Hitchcock et al., 1959; Sudworth, 1908). In California, F. latifolia is found in the North Coast, Klamath, and Cascade Ranges, the Sacramento and San Joaquin Valleys, and the foothills of the Sierra Nevada (Fig. 3). (Griffin & Critchfield, 1972). The distributions of the fossil counterparts of F. latifolia resemble to some extent those of the previous two species. Fraxinus coulteri was relatively common on the Columbia Plateau and in border areas in the Miocene (Chaney & Axelrod, 1959) and extended southward in Nevada to the region of the Middlegate and Eastgate floras (Axelrod, 1956 and unpublished). F. caudata, a probable descendent of F. coulteri (Chaney & Axelrod, 1959), was common in central California during the Pliocene (see Dorf, 1930; Axelrod, 1944b, d) when its distribution reached northward at least as far as the Troutdale flora of northwestern Oregon (Chaney, 1944b).

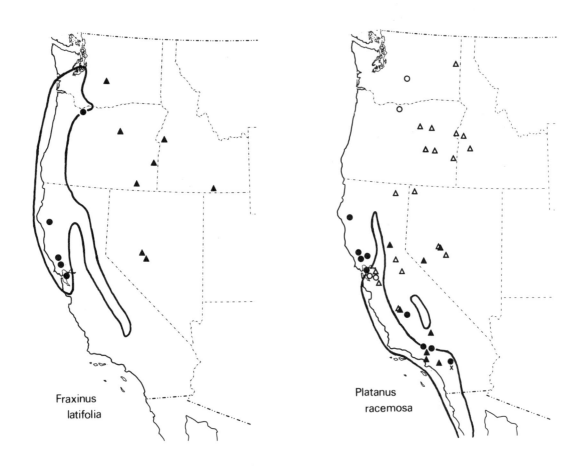

Figure 3. Past and present geographical distributions of Fraxinus latifolia and
Platanus racemosa. The symbols are the same as in Fig. 2. For Platanus
racemosa, the open symbols are for P. dissecta and the closed symbols are
for P. paucidentata.

The California sycamore, Platanus racemosa, ranges in distribution from the upper
reaches of the Sacramento River southward into Baja California (Fig. 3) (Griffin &
Critchfield, 1972; Little, 1976). In the Central Valley, this species is locally abundant
along the Sacramento and San Joaquin Rivers, ascending their main tributaries to low elevations in the Sierran foothills. It is notably absent from the North Coast Ranges and
the western side of the Sacramento Valley (Jepson, 1910). It is distributed throughout
the South Coast Ranges, where it is "one of the most widely distributed arboreous species"
(Jepson, 1910) and occurs in the Transverse and Peninsular Ranges of southern California
(Griffin & Critchfield, 1972). Two distinct late Tertiary species have been referred to
the modern P. racemosa. To the north, Platanus dissecta is a characteristic species in the
Miocene floras of the Columbia Plateau and northern Great Basin (Chaney & Axelrod, 1959).
It survived into the Pliocene in this region as evidenced by the Dalles flora of Oregon
(Chaney, 1944a) and the Upper Ellensburg flora of Washington (Smiley, 1963). To the south,
Platanus paucidentata is a characteristic species in both the Miocene and Pliocene floras
of southern California (Axelrod, 1939, 1940, 1950c, d). The distributions of these two
species overlapped in central Nevada in the Miocene (Axelrod, 1956) and in central California in the Miocene and Pliocene (see Axelrod, 1944a, b; Renney, 1972). The question
arises as to which of these species is more closely allied to the modern P. racemosa.
Judging from the available record, it appears that P. paucidentata shows more definite
relationships to the modern species (Axelrod, 1939, 1956, 1976a), while P. dissecta may be
more nearly related to the modern P. orientalis of the Middle East or P. occidentalis of
the eastern U.S. (Axelrod, 1956; Renney, 1972). The two fossil species probably diverged
from a common ancestor during the early or middle Tertiary. Leaves of the modern P.
racemosa appear in abundance in the Pleistocene Soboba flora of southern California
(Axelrod, 1966).

Populus fremontii, the Fremont cottonwood, is distributed throughout the Southwest, extending from California eastward to Colorado and Texas and southward into Mexico (Little, 1971). The species occurs throughout California (Fig. 4). (Griffin & Critchfield, 1972) and is most abundant in the San Joaquin and Sacramento Valleys. The fossil counterpart, Populus prefremontii, was relatively common in southern and central California in the Miocene and Pliocene (see Axelrod, 1939, 1944b, 1950c), but is essentially unrecorded from most of the Great Basin and regions to the north in the late Tertiary. (Populus russelli, which shows affinity to both P. deltoides of the eastern U.S. and P. fremontii of the western U.S., is recorded from the Ellensburg flora of Washington (Smiley, 1963). Two of the specimens in the flora of the Tyrell locality resemble P. prefremontii (Smiley, 1963)). A leaf impression of the modern species appears in the Pleistocene Soboba flora (Axelrod, 1966).

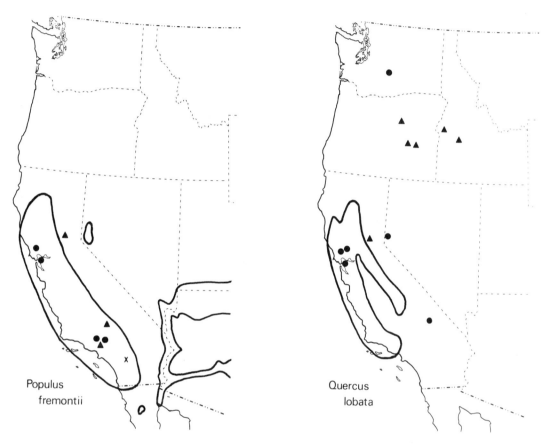

Figure 4. Past and present geographical distributions of Populus fremontii and Quercus lobata. Symbols as in Fig. 2.

The valley oak, Quercus lobata, is endemic to California (Little, 1971) where it occurs throughout the Central Valley, the Sierran foothills, and the Coast and Transverse Ranges (Fig. 4) (Griffin & Critchfield, 1972). Quercus prelobata, the fossil counterpart, is recorded from several of the Miocene floras of Oregon and adjacent Idaho (Chaney & Axelrod, 1959) and from the Miocene Remington Hill flora of central California (Condit, 1944a). It also appears in the Pliocene Upper Ellensburg flora of Washington (Smiley, 1962), Verdi flora of Nevada (Axelrod, 1958), and Napa flora of central California (Axelrod, 1950a). A second fossil species, Quercus moragensis, has also been referred to Q. lobata. This species appears in the middle Pliocene Mulholland and Petaluma floras of central California (Axelrod, 1944b, d) and differs from Q. prelobata by its consistently smaller size. Pollen similar to the modern Q. lobata is recorded from several late Pliocene pollen floras from the central and southern Sierra Nevada (Axelrod & Ting, 1960).

The modern distribution of Salix lasiandra, the yellow willow, extends from Alaska and the Yukon southward to California and New Mexico, though it is largely or wholly absent from the dry interior country of eastern Washington and Oregon, southern Idaho, Nevada, and Utah (Hitchcock et al., 1964; Little, 1976). In California, the species is widely distributed in the Coast Ranges, Central Valley, Sierra Nevada (particularly the foothills), and southern California (outside of the deserts) (Fig. 5) (Jepson, 1910). Its close fossil ally, Salix hesperia, was widely distributed in the western U. S. in the later Tertiary. Its Miocene distribution extended from central California in the Neroly and Remington Hill floras (Condit, 1938, 1944a) and central Nevada in the Middlegate flora (Axelrod, 1956 and unpublished) northward into Idaho, Oregon, and Washington (see Axelrod, 1964; Chaney & Axelrod, 1959; Smiley, 1963). In the Pliocene, this species ranged from southern California and Oregon to Washington (see Axelrod, 1944b, d; Chaney, 1944a; Smiley, 1963). As with Alnus rhombifolia, pollen resembling that of the modern Salix lasiandra is recorded from several late Pliocene and early Pleistocene pollen floras from the southern Sierra Nevada (Axelrod & Ting, 1960, 1961).

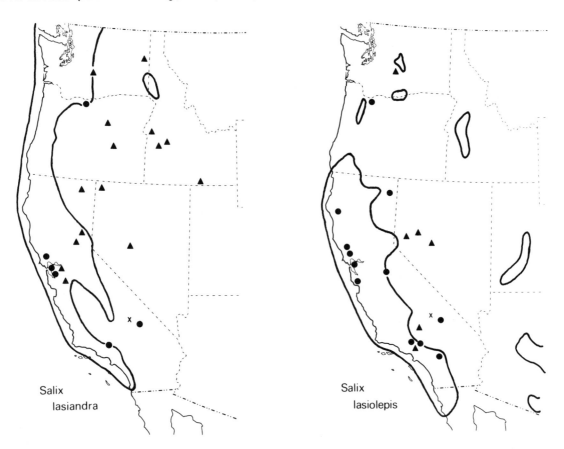

Figure 5. Past and present geographical distributions of Salix lasiandra and Salix lasiolepis. Symbols as in Fig. 2.

Salix lasiolepis, the arroyo willow, ranges from Baja California eastward to west Texas and adjacent Mexico and northward to southern British Columbia (Fig. 5) (Hitchcock et al., 1964; Little, 1976). In contrast to the yellow willow, this species occurs only to the east of the Cascades in Oregon and Washington and is common in the dry interior country of western Idaho, Nevada, and southern Utah (Hitchcock et al., 1964). In California, the distribution of S. lasiolepis is similar to that of S. lasiandra, though the former is more common than the latter in the Central Valley and occurs occasionally in the desert (Jepson, 1910; Munz, 1959). Salix wildcatensis, the fossil counterpart of S. lasiolepis, is recorded from the Miocene Tehachapi and Mint Canyon floras of southern California and the Middlegate, Purple Mountain, and Chloropagus floras of central Nevada (Axelrod, 1939, 1940, 1956, 1976a, and unpublished). A single leaf impression similar to that of S. wildcatensis is also recorded from the late Miocene Lower Ellensburg flora of Washington (Smiley, 1963). In the Pliocene, this species was common in southern and central California (see Axelrod, 1944b, d, e, 1950b, c, d) and ranged at least as far northward as the Troutdale flora of Oregon (Chaney, 1944b). As with S. lasiandra, pollen similar to the modern S. lasiolepis is recorded from several late Pliocene and early Pleistocene pollen floras from the southern and central Sierra Nevada (Axelrod & Ting, 1960, 1961).

That no two of these species have identical past or present geographical distributions reflects their independent evolution and migration in response to changing environments. The general similarity of the distributions of certain of the species, though, is sufficient to suggest two primary floristic sources for the modern community. Species such as Acer (negundo), Alnus (rhombifolia), Fraxinus (latifolia), Quercus (lobata), and Salix (lasiandra) are recorded primarily in the north, where they apparently contributed to a mixed mesophytic hardwood-conifer forest (the "Arcto-Tertiary Geoflora") that blanketed much of the northern U. S. and Canada during the middle and late Tertiary. In contrast, species such as Platanus (racemosa), Populus (fremontii), and Salix (lasiolepis) are recorded primarily in the south, where they apparently contributed to a sub-arid woodland and scrub vegetation (the "Madro-Tertiary Geoflora") that spread outward from this region with expanding dry climate in the middle and late Tertiary (Axelrod, 1958). (The modern species epithets appear in parentheses in this and subsequent discussions whenever the reference is actually to the ancestors of the modern taxa. This seems preferable to using the less familiar fossil names).

Within this larger framework of general similarities, the different distribution patterns reveal the somewhat different responses of the individual species to changing environments during the later Tertiary. As the climate became cooler, drier, and less equable (Axelrod, 1968), certain of the northern species, such as Acer (negundo var. Californicum) and Quercus (lobata), were apparently eliminated from the northern part of their ranges and became restricted to California. Other species, such as Fraxinus (latifolia) and Salix (lasiandra), remained in the north but gradually were confined to the relatively mild coastal strip where the effects of changing climate were somewhat diminished. Still other species, such as Alnus (rhombifolia), were apparently able to adapt to the new climatic regimes and survive in relatively unmodified form in the northern interior regions. Of the species that were restricted coastward and southward, some gave rise to derivative forms that adapted to the new climate in the interior, and survive in modern form as varieties (Acer negundo var. interior) or closely related species (Salix caudata) of counterparts to the west (Chaney & Axelrod, 1959). The differentiation of S. caudata apparently occurred relatively early. A fossil species, Salix venosiuscula, is recorded from the Miocene Thorn Creek and Lower Idaho floras that shows relationship to the modern S. caudata (Chaney & Axelrod, 1959). Though some authors consider S. caudata to be a variety of S. lasiandra (Hitchcock, et al., 1964), its appearance in the Miocene record suggests specific status. A similar segregation of derivative forms apparently occurred in several of the southern species. As Platanus (racemosa) gradually became restricted to California and adjacent Mexico in response to increasing aridity throughout the southwest, the closely related P. wrightii of Arizona was apparently differentiated (Axelrod, 1938). In a similar manner, Populus (fremontii) gave rise to the complex of varieties (and species) that now range across the Southwest from California to Texas (Axelrod, 1939). In contrast, Salix (lasiolepis) appears not only to have survived in relatively unmodified form in the Southwest and California, but also to have extended its range northward to British Columbia. Its occurrence only in the dry interior regions in the north and not along the coastal strip reflects its southern origins. In relation to the origins of the modern riparian forest, all of these species are recorded in California at least by the middle Pliocene, with the large majority being recorded by the later Miocene.

Table 2. Approximate ages of selected fossil floras from the western United States that contain ancestors of California's lowland riparian taxa (after Axelrod, 1972, 1977).[1]

Age (million years ago)	Southern California	Central California	Nevada	Washington, Oregon, Idaho
Quarternary	Soboba			
		Sonoma		
		Napa		
Pliocene	Anaverde	Petaluma	Verdi	
	Mt. Eden	Mulholland		
	Piru Gorge		Truckee	Troutdale
				Dalles
– 10 -------		Remington Hill		Ellensburg
	Mint Canyon	Table Mountain	Aldrich Station	
		Neroly	Fallon	Stinking Water
			Chloropagus	
				Trapper Creek
				Blue Mountain
Miocene		Temblor		Mascall
				Succor Creek
			Middlegate	
	Tehachapi		Eastgate	
– 20				

[1]The vertical bars connect fossil floras of similar age.

The history of these individual lineages provides the background for understanding the history of the modern community. In addition, it is important to recall the major climatic changes that occurred in California and the western United States during the later Tertiary that had a major impact on the composition and distribution of the ancient communities: the general trend to cooler and drier climate, to a shift in the seasonal distribution of rainfall, and to progressively decreasing equability (Axelrod, 1968). The plant communities of 20 millions ago lived under significantly milder winters and summers with a more even seasonal distribution of precipitation than do the modern communities.

An important consideration in any discussion of community history is the problem of assigning individualistic species of definite communities. This difficulty is amplified in the fossil record since we have little reason to assume that the habitat requirements of lineages have not changed with time. It is possible within certain limits, though, to recognize general assemblages of species that are repeated in the modern landscape. In addition, although individual lineages may have had their tolerances altered during geologic time, there is little likelihood that every member of an assemblage will have changed in precisely the same manner (Condit, 1938; Whittaker, 1975). Therefore, when an association of species in a fossil flora resembles in composition a modern community, we may speculate that the fossil species formed a community in the ancient landscape with generally the same major habitat requirements as the modern counterpart.

Southern California

The middle Miocene Tehachapi flora of interior southern California (Axelrod, 1939) is dominated by species whose modern relatives are typical of the oak woodland, semi-desert scrub, and arid subtropical scrub associations of southern California, the southwestern U. S., and Mexico. Several species are also represented in the flora that may have contributed to a rich riparian community along streams in the region. These include Platanus (racemosa), Populus (fremontii), and Salix (lasiolepis), together with species of Acer, Celtis, Erythrea, Fraxinus, Persea, Populus, and Sabal whose modern relatives are now in the summer-wet region of the southwestern U. S. and adjacent Mexico. The overall composition of the Tehachapi flora suggests that species lived under a climate of mild, frostless winter and ample summer rainfall (Axelrod, 1972).

The rich late Miocene Mint Canyon flora (Axelrod, 1940) is also dominated by oak savanna and thorn shrub species, and includes many of the riparian taxa found in the Tehachapi flora. Platanus (racemosa), Populus (fremontii), and Salix (lasiolepis) are represented, as are exotic species of Celtis, Juglans, Lyonothamnus, and Persea. The general composition of this flora suggests climatic conditions generally similar to those of the Tehachapi flora, though slightly more arid (Axelrod, 1972).

This general trend toward increasing aridity and decreasing summer precipitation is well-reflected in the composition of the middle Pliocene Mt. Eden, Anaverde, and Piru Gorge floras of southern California (Axelrod, 1950b, e, d). Oak woodland-savanna and associated chaparral species dominate these floras. Decreased summer rainfall and lower winter temperatures are suggested by the poor numerical representation relative to the Miocene floras of species allied to modern members of the woodland and scrub vegetation of northern Mexico. The riparian vegetation represented in these floras is still relatively rich in these exotic species, suggesting that the effect of changing climate was somewhat diminished and relict survival was facilitated in sites with abundant water. (Similar situations are not uncommon in the modern landscape, where riparian vegetation is often zonal in character.) Possible riparian taxa in these floras include Juglans (californica), Platanus (racemosa), Populus (fremontii), Salix (goodingii), S. (lasiandra), S. (lasiolepis), and exotic species of Acer, Juglans, Persea, Sapindus, and Sabal.

The disappearance of the various exotic species in southern California as a consequence of decreased summer rainfall was essentially complete by the early Pleistocene as evidence by the composition of the Soboba flora (Axelrod, 1966). Only two species in the flora, Acer (brachypterum) and Magnolia (grandiflora), no longer occur in California. Its riparian woodland is essentially modern in aspect, including Acer (negundo), Cornus (californica), Platanus (racemosa), Populus (fremontii), Salix (laevigata), and Toxicodendron (diversiloba). The Acer and Magnolia species probably contributed to this community. Species of yellow pine forest, bigcone spruce forest, and chaparral are also represented in the flora.

Central California

The middle Miocene Temblor flora of central California (Renney, 1972) is composed of various deciduous hardwood and broadleaved evergreen species whose nearest relatives are found in the western U. S., the eastern U. S., and eastern Asia. Modern and geologic relationships suggest that several of the species may have occupied estuarine and river-border sites (Renney, 1972). Platanus (racemosa) and Cornus (californica) are represented in the flora, as are species of Carya, Diospyros, Glyptostrobus, Magnolia, Nyssa, Persea, Platanus, and Quercus with modern relatives in lowland riparian habitats in the summer-wet regions of the eastern U. S. and eastern Asia. The composition of the Temblor flora suggests a climate with mild summer and winter temperatures and an even seasonal distribution of abundant precipitation.

The late Miocene Remington Hill and Table Mountain floras of the western Sierran slope (Condit, 1944a, b) are also dominated by deciduous hardwood forest and mixed-evergreen forest species, and include possible riparian taxa such as Acer (negundo), Juglans (californica), Platanus (racemosa), Populus (fremontii), Quercus (lobata), Salix (lasiandra), and exotic species of Aesculus, Carya, Cercis, Liquidambar, Magnolia, Nyssa, Persea, Quercus, Rhododendron, and Ulmus. The increased proportion of oak woodland and chaparral species of southern affinity in these floras relative to the Temblor flora reflects the general trend to drier climate and the northward shift of taxa of Madro-Tertiary affinity.

The effects of increasing aridity were still somewhat diminished over the lowlands of west-central California at this time as judged from the composition of the Neroly flora (Condit, 1938). This flora is composed primarily of Taxodium-swamp and floodplain species whose modern relatives occur in the mild, summer-rain regions of the southwestern U. S. Among the possible riparian taxa are Alnus (rhombifolia), Salix (lasiandra), and species of Magnolia, Nyssa, Persea, Platanus, Populus, and Prunus that no longer occur in the western U. S.

It is during the middle Pliocene that the late Tertiary trend toward increasing aridity reached its peak and oak woodland-savanna and associated chaparral species became dominant throughout most of central California (Axelrod, 1948, 1972). This is evidenced by the rich middle Pliocene Mulholland flora of west-central California (Axelrod, 1944b), which contains numerous species whose modern relatives contribute to oak woodland, chaparral, and mixed-evergreen forest in California. Also included in the flora (and contributing most of the specimens) are ancestors of most of the dominants of the modern riparian forest of the Sacramento River: Alnus (rhombifolia), Fraxinus (latifolia), Platanus (racemosa), Populus (fremontii), Quercus (lobata), Salix (lasiandra), S. (lasiolepis), and Toxicodendron (diversiloba). All of these species (except poison oak) are also represented in the middle Pliocene Petaluma flora of the same region (Axelrod, 1944d), suggesting that the modern forest was rather well-differentiated by this time. A small number of exotic species still contributed to the riparian community, including those now restricted to the southwestern U. S. (Celtis, Populus, and Sapindus) and the eastern U. S. (Nyssa, Populus and Ulmus). The latter species indicate that some summer rainfall was still present at this time (Axelrod, 1972).

The composition of the Sonoma and Napa floras of west-central California (Axelrod, 1944d, 1950a) reflects an increase in annual rainfall over that of the middle Pliocene (Axelrod, 1972). Mesophytic conifer and hardwood species are more numerous in these floras than in the older Mulholland and Petaluma floras of the same region, though oak woodland taxa are still relatively abundant. River-border and flodplain sites were apparently occupied by Alnus (rhombifolia), Fraxinus (latifolia), Platanus (racemosa), Quercus (lobata), Salix (laevigata), S. (lasiandra), Smilax (californica), and a few exotic species of Persea and Ulmus.

Western Nevada

The middle Miocene Middlegate and Eastgate floras of west-central Nevada (Axelrod, 1956 and unpublished) include both conifer and deciduous hardwood species of northern affinity and evergreen sclerophyllous species of southern affinity. This mixture of floristic sources is evident in the possible riparian species represented in the floras: Acer (negundo), Fraxinus (latifolia), Platanus (racemosa), Salix (lasiandra), S. (lasiolepis), and several exotic species from the eastern U. S. (Betula) and the southwestern U. S. (Fraxinus and Persea) are recorded. As in California at this time, the composition of these middle Miocene floras suggests relatively mild temperatures and some summer rainfall.

The composition of the later Miocene floras from the lowlands still reflects this mixture of floristic sources, though there is a decrease in overall species diversity. This presumably represents a response to spreading aridity and decreasing summer rainfall, and is reflected to some extent in the composition of the riparian element of these floras. Ancestors of the modern riparian taxa of the Sacramento River are recorded only infrequently. The composition of the Aldrich Station, Chloropagus, and Fallon floras (Axelrod, 1956) suggests that lowland stream and lake-border sites may have been occupied by Platanus (racemosa), Salix (lasiolepis), and exotic species of Bumelia, Populus, and Salix.

By the middle Pliocene, conifer-hardwood and mixed-evergreen forests had become restricted to the east Sierran slope, and oak woodland-savanna species dominated the lowlands (Axelrod, 1972). Taxa that contributed the well-developed floodplain forests in central California at this time are essentially unrecorded in western Nevada. Julgans (californica) appears in the Truckee Flora (Axelrod, 1950b) and a single specimen of Quercus (lobata) is recorded from the Verdi flora (Axelrod, 1958a) of this region.

Oregon-Washington-Idaho

The composition of the middle Miocene Blue Mountains, Mascall, Succor Creek, and Trapper Creek floras (Chaney, 1959; Axelrod, 1964) suggest that lower elevations in the Columbia Plateau and adjacent regions supported a rich Taxodium-swamp association that was replaced at higher levels and in the interior by mixed deciduous hardwood forest and montane conifer-hardwood forest. Most taxa in these floras no longer survive in the western U. S. Modern relationships suggest that a number of species may have occupied stream and lake-border sites. Acer (negundo), Alnus (rhombifolia), Fraxinus (latifolia), Quercus (lobata), Salix (lasiandra), and Smilax (californica) are recorded in these floras, as are species of Acer, Carya, Celtis, Diospyros, Liquidamber, Nyssa, Persea, Platanus, Populus, Quercus, and Ulmus whose modern relatives occupy lowland riparian sites in the eastern U. S. and eastern Asia. Under a mild, moist Miocene climate, these species were probably less restricted to river-border sites than are the modern lowland riparian species in California. They probably also contributed to rich upland hardwood forests much as their modern relatives do today in the eastern U. S.

This rich riparian forest also appears in the late Miocene Stinking Water flora of Oregon (Chaney, 1959), where Acer (negundo), Alnus (rhombifolia), Quercus (lobata), Salix (lasiandra), Smilax (californica), and exotic species of Acer, Glyptostrobus, Platanus, Populus, and Ulmus are recorded. The large number of exotics in the flora suggests that summer rainfall was present in larger amount than to the south.

As in California and Nevada, fewer exotic species are recorded in the Pliocene floras of Oregon, Washington, and Idaho, an apparent consequence of cooling and drying climate and decreasing summer rainfall. The composition of the early Pliocene Troutdale, Upper Ellensburg, and Dalles floras (Chaney, 1944a, b; Smiley, 1963) suggests that riparian sites may still have supported a relatively rich community, however. Acer (negundo), Cornus (californica), Fraxinus (latifolia), Salix (lasiandra), S. (lasiolepis), and exotic species of Cercis, Diospyros, Liquidambar, Platanus, Tilia, and Ulmus occur in these floras.

Discussion

The evidence from these fossil floras suggests that lowland riparian forests comparable to that along the modern Sacramento River have had a long and nearly continuous history in the western United States during the last 20 million years. The available evidence also suggests that the modern forests are impoverished representatives of richer, more generalized ancestral communities that included taxa no longer found in California (Table 3). These widespread ancestral communities showed regional variation as a consequence of major climatic differences from north to south. In southern regions, the riparian communities originally included several species with relatives in the modern forest (P. racemosa, P. fremontii, and S. lasiolepis) plus numerous taxa now restricted to the summer-wet region of the southwestern U. S. and northern Mexico. In contrast, the original riparian communities of northern regions included several other species with relatives in the modern forest (A. negundo, A. rhombifolia, F. latifolia, Q. lobata, and S. lasiandra) plus many others now confined to the summer-wet regions of the eastern U. S. and eastern Asia. It is in the intermediate areas that we first see the intermingling of these northern and southern riparian taxa that is apparent in the modern community. This is first evident in the interior (Middlegate), where the northward migration of southern taxa with spreading aridity was apparently aided by the Sierra Nevada rain shadow. This mixed type of community

Table 3. Some modern species from the eastern U.S., eastern Asia, the southwestern U.S., and northern Mexico with counterparts in the late Tertiary fossil record of the western United States. The ancestors of these species may have occupied riparian and floodplain sites with the species listed in Table 1.

Eastern U.S.	Eastern Asia
Acer saccharinum	*Acer pictum*
Carya ovata	*Glyptostrobus pensilis*
Celtis mississippiensis	*Zelkova serrata*
Diospyros virginiana	Southwestern U.S.
Fraxinus caroliniana	*Celtis reticulata*
Liquidambar styraciflua	*Juglans major*
Magnolia grandiflora	*Populus angustifolia*
Nyssa aquatica	*Sapindus drummondii*
Persea borbonia	Northern Mexico
Platanus occidentalis	*Persea podadenia*
Populus heterophylla	*Populus brandegeei*
Quercus bicolor	*Sabal uresana*
Taxodium distichum	

subsequently appeared on the western slopes of the Sierra Nevada (Remington Hill) and disappeared from western Nevada. It had become well-established over lowland west-central California by the middle Pliocene (Mulholland) and persisted in this region with modifications down to the present. In all of these regions, we see the gradual loss of the exotic taxa in the communities as the climate became progressively cooler, drier, and less equable, and as summer rainfall was reduced.

Summary

The history of California's lowland riparian community follows the principles outlined by Axelrod (1977) for community evolution in general in the western United States during the late Tertiary:

(1) The modern riparian community includes taxa of different floristic sources. Acer negundo, Alnus rhombifolia, Fraxinus latifolia, Quercus lobata, and Salix lasiandra represent an "Arcto-Tertiary" or northern element, while Platanus racemosa, Populus fremontii, and Salix lasiolepis represent a "Madro-Tertiary" or southern element.

(2) The modern community is a relatively impoverished representative of richer, more generalized ancestral communities that included taxa related to modern species found only in the southwestern U. S. and northern Mexico (species of Acer, Celtis, Juglans, Persea, and Sapindus), the eastern U. S. (species of Carya, Liquidambar, Magnolia, Nyssa, and Ulmus), and eastern Asia (species of Glyptostrobus, Ulmus, and Zelkova). These exotic taxa were eliminated from the ancestral communities by the major climatic changes of the later Tertiary.

(3) Several of the dominant species in the modern community have apparently been associated, as ancestral forms in ancient communities, for a time span of nearly 20 million years.

Epilogue

We see that lowland riparian forests have had a long evolutionary history in California. We also see that we are on the verge of destroying these magnificent forests. For the natural processes that have shaped the modern community to continue, we need to establish a system of healthy and extensive riparian preserves throughout the entire state. It would be nice if we could insure the existence of such a system for another 20 million years. If not, then for the next 100 years would be a good starting point.

Acknowledgements

The ideas expressed in this assay stem directly and entirely from conversations with Professor D. I. Axelrod, Department of Botany, University of California at Davis. While he may not agree with my interpretation of his writings, I wish to thank him for his unlimited generosity with his time and his ideas.

I wish to thank Drs. D. I. Axelrod, J. Major, and R. W. Peacey for reading the manuscript.

Literature Cited

Axelrod, D. I. 1937. A Pliocene flora from the Mount Eden beds, southern California. Carnegie Inst. Wash. Pub. 476:125-183.

_____. 1939. A Miocene flora from the western border of the Mohave Desert. Carnegie Inst. Wash. Pub. 516:1-129.

_____. 1940. The Mint Canyon flora of southern California: a preliminary statement. Amer. Jour. Sci. 238:577-585.

_____. 1944a. The Black Hawk Ranch flora. Carnegie Inst. Wash. Pub. 553:91-101.

_____. 1944b. The Mulholland flora. Carnegie Inst. Wash. Pub. 553:103-145.

_____. 1944c. The Oakdale flora. Carnegie Inst. Wash. Pub. 553:147-165.

_____. 1944d. The Sonoma flora. Carnegie Inst. Wash. Pub. 553:167-206.

_____. 1944e. The Pliocene sequence in California. Carnegie Inst. Wash. Pub. 553:207-224.

_____. 1944f. The Alturas Flora. Carnegie Inst. Wash. Pub. 553:

_____. 1948. Climate and evolution in western North America during Middle Pliocene time. Evolution 2:127-144.

_____. 1950a. A Sonoma florule from Napa, California. Carnegie Inst. Wash. Pub. 509:23-71.

_____. 1950b. Further studies of the Mount Eden flora, southern California. Carnegie Inst. Wash. Pub. 590:73-117.

_____. 1950c. The Anaverde flora of southern California. Carnegie Inst. Wash. Pub. 590:119-158.

_____. 1950d. The Piru Gorge flora of southern California. Carnegie Inst. Wash. Pub. 590:159-214.

_____. 1950e. Evolution of desert vegetation. Carnegie Inst. Wash. Pub. 590:215-306.

_____. 1956. Mio-Pliocene floras from west-central Nevada. Univ. Calif. Publ. Geol. Sci. 33:1-316.

_____. 1958a. The Pliocene Verdi flora of western Nevada. Univ. Calif. Pub. Geol. Sci. 34:91-160.

Axelrod, D. I. 1958b. Evolution of the Madro-Tertiary Geoflora. Bot. Rev. 24:433-509.

_____. 1959. Late Cenozoic evolution of the Sierran bigtree forest. Evolution 13:9-23.

_____. 1962. A Pliocene Sequoiadendron forest from western Nevada. Univ. Calif. Publ. Geol. Sci. 39:195-268.

_____. 1964. The Miocene Trapper Creek flora of southern Idaho. Univ. Calif. Publ. Geol. Sci. 51:1-180.

_____. 1966. The Pleistocene Soboba flora of southern California. Univ. Calif. Publ. Geol. Sci. 60:1-109.

_____. 1967a. Evolution of the Californian closed-cone pine forest. In: Proceedings of the Symposium on the Biology of the California Islands. (Ed. by R. N. Philbrick), pp. 93-150. Santa Barbara Botanical Garden, Santa Barbara.

_____. 1967b. Geologic history of the California insular flora. In: Proceedings of the Symposium on the Biology of the California Islands (Ed. by R. N. Philbrick), pp. 267-316. Santa Barbara Botanical Garden, Santa Barbara.

_____. 1968. Developments, trends, and outlooks in paleontology. Late Tertiary plants (Oligocene-Pliocene). J. Paleont. 42:1358.

_____. 1972. History of the Mediterranean ecosystem in California. In: Mediterranean-Type Ecosystems, Origin and Structure (Ed. by F. di Castri & H. A. Mooney), pp. 225-277. Springer, Berlin.

_____. 1975. Evolution and biogeography of Madrean-Tethyan sclerophyl vegetation. Ann. Mo. Bot. Gard. 62:280-334.

_____. 1976a. Evolution of the Santa Lucia fir (Abies bracteata) ecosystem. Ann. Mo. Bot. Gard. 63:24-41.

_____. 1976b. History of the coniferous forests, California and Nevada. Univ. Calif. Publ. Bot. 70:1-62.

_____. 1977. Outline history of California vegetation. In: Terrestrial Vegetation of California (Ed. by M.G. Barbour & J. Major), pp. 139-193. Wiley-Interscience, New York.

Axelrod, D. I., and W. S. Ting. 1960. Late Pliocene floras east of the Sierra Nevada. Univ. Calif. Publ. Geol. Sci. 39:1-118.

_____ and _____. 1961. Early Pleistocene floras from the Chagoopa surface, southern Sierra Nevada. Univ. Calif. Publ. Geol. Sci. 39:119-194.

Chaney, R. W. 1938. The Deschutes flora of eastern Oregon. Carnegie Inst. Wash. Pub. 476:185-216.

_____. 1944a. The Dalles flora. Carnegie Inst. Wash. Pub. 553:285-321.

_____. 1944b. The Troutdale flora. Carnegie Inst. Wash. Pub. 553:323-351.

_____. 1959. Miocene floras of the Columbia Plateau. I. Composition and interpretation. Carnegie Inst. Wash. Pub. 617:1-134.

Chaney, R. W., and D. I. Axelrod. 1959. Miocene floras of the Columbia Plateau. II. Systematic considerations. Carnegie Inst. Wash. Pub. 617:135-237.

Condit, C. 1938. The San Pablo flora of west-central California. Carnegie Inst. Wash. Pub. 476:217-268.

_____. 1944a. The Remington Hill flora. Carnegie Inst. Wash. Pub. 553:21-55.

Condit, C. 1944b. The Table Mountain flora. Carnegie Inst. Wash. Pub. 553:57-90.

Dorf, E. 1933. Pliocene floras of California. Carnegie Inst. Wash. Pub. 412:1-112.
_____. 1936. A late Tertiary flora from southwestern Idaho. Carnegie Inst. Wash. Pub. 476:73-124.

Griffin, J. R., and W. B. Critchfield. 1972. The Distribution of Forest Trees in California. U. S. Dept. of Agriculture, Berkeley.

Hitchcock, C. L., A. Conquist, M. Ownbey, and J. W. Thompson. 1959. Vascular Plants of the Pacific Northwest. Part. 4. Ericaceae through Campanulaceae. University of Washington Press, Seattle.

_____, _____, _____, and _____. 1964. Vascular Plants of the Pacific Northwest. Part 2. Salicaceae to Saxifrigageae. University of Washington Press, Seattle.

Jepson, W. L. 1910. The Silva of California. The University Press, Berkeley.

Little, E. L. 1971. Atlas of United States Trees. Vol. 1. Conifers and Important Hardwoods. U. S. Dept. of Agriculture, Washington, D. C.

_____. 1976. Atlas of United States Trees. Vol. 3. Minor Western Hardwoods, U. S. Dept. of Agriculture, Washington, D. C.

Munz, P. A. 1959. A California Flora. University of California Press, Berkeley.

Renney, K. M. 1972. The Miocene Temblor flora of west-central California. M.Sc. Thesis, University of California, Davis.

Smiley, C. J. 1963. The Ellensburg flora of Washington. Univ. Calif. Publ. Geol. Sci. 35:157-275.

Smith, H. V. 1941. A Miocene flora from Thorn Creek, Idaho. Amer. Midl. Natur. 25:473-522.

Sudworth, G. B. 1908. Forest Trees of the Pacific Slope. U. S. Dept. of Agriculture, Washington, D. C.

_____. 1934. Poplar, Principal Tree Willows, and Walnuts of the Rocky Mountain Region. U. S. Dept. of Agriculture, Washington, D. C.

Whittaker, R. H. 1975. Communities and Ecosystems. MacMillan Publ. Co., New York.

Chapter 4

RIPARIAN FORESTS OF THE SACRAMENTO VALLEY, CALIFORNIA

Kenneth Thompson
Department of Geography
University of California, Davis

Although edaphic and biotic influences precluded trees from most of the Sacramento Valley in its pristine condition, the riparian lands (mainly natural levees) supported a flourishing tree growth-valley oak, sycamore, cottonwood, willow, and other species. A number of factors contributed to their presence - principally sub-irrigation, fertile alluvial loam soils, and relative freedom from surface waterlogging and fire. These riparian forests varied considerably in width, from a narrow strip to several miles. They also varied greatly in the spacing of the trees, from irregular open to fairly crowded stands, but were generally of sufficient extent and closeness to justify the term "forest".

Pristine Condition of the Riparian Lands

Among the first outsiders to visit the Sacramento Valley were fur trappers of the Hudson's Bay Company in the period prior to 1814. The Spaniard Luis Antonio Arguello investigated the valley in 1817 and again in 1821, and Jedediah Smith, in 1825, may have been the first American to reach the Sacramento River. However, it was not until the 1840's that significant outside influence was felt in the northern end of the Central Valley. This seclusion, however, could not survive the meteoric developments of the Americanization of California. After 1849 came a huge influx of population, lured by gold but often quick to adopt other pursuits. These immigrants, mostly with rural backgrounds, could not overlook the agricultural promise of the Sacramento Valley. Heightening the attractions of the Sacramento Valley for agricultural settlement was its virtually vacant condition. Its relatively sparse and peaceful aboriginal population, having been greatly reduced in numbers by an epidemic in the early 1830's, was unable to offer more than token resistance to the American invaders (Cook, 1955).

After recognizing the promise of the Sacramento Valley, the invading Americans quickly began its realization. To do this called for new patterns of occupance and land use; and, in the initiation of these, the environment was substantially modified. The agencies of change were sufficiently drastic to transform the physical, biotic, and cultural landscape. One of the very first transformations concerned the natural levees and riparian lands, which were thickly forested in their pristine condition.

Because of the brief period between initial investigation and development, little information was accumulated on the aboriginal condition of the Sacramento Valley. One of the earliest observers to report on the riparian forests was John Work (in Mahoney, 1945), in the course of a fur-trapping expedition from his headquarters at Fort Vancouver. Writing in 1832, he described the riparian forests of the Sacramento Valley below Red Bluff as follows:

> All the way along the river here there is a belt of woods (principally oak) which is surrounded by a plain with tufts of wood here and there which extend to the foot of the mountain, where the hills are again wooded.

Another early visitor to the Sacramento Valley, Captain Sir Edward Belcher, R. N., noted the profusion of oak, ash, plane, laurel, sumach (sic), hiccory (sic), walnut, roses, wild grapes, arbutus, and other small shrubs in the vicinity of the river (Belcher 1843). He described its lower course as follows:

> Having entered the Sacramento, we soon found that it increased in width as we advanced, and at our noon station of the second day was about one-third of a mile wide. The marshy land now gave way to firm ground, preserving its level in a most remarkable manner, succeeded by banks well wooded with oak, planes, ash, willow, chestnut (sic), walnut, poplar, and brushwood. Wild grapes in great abundance overhung the lower trees, clustering to the river, at times completely overpowering the trees on which they climbed, and producing beautiful varieties of tint. . . . Our course lay between banks. . . . These were, for the

35

most part, belted with willow, ash, oak, or plane (Platanus racemosa), which latter, of immense size, overhung the stream, without apparently a sufficient hold in the soil to support them, so much had the force of the stream denuded their roots.

Within, and at the very verge of the banks, oaks of immense size were plentiful. These appeared to form a band on each side, about three hundred yards in depth, and within (on the immense park-like extent, which we generally explored when landing for positions) they were seen to be disposed in clumps, which served to relieve the eye, wandering over what might otherwise be described as one level plain or sea of grass. Several of these oaks were examined, and some of the small felled. The two most remarkable measured respectively twenty-seven feet and nineteen feet in circumference, at three feet above ground. The latter rose perpendicularly at a (computed) height of sixty feet before expanding its branches, and was truly a noble sight.

Most of the historical reports give no indication of the actual depth of the woodland. Where Belcher examined the lower Sacramento banks, probably the delta section, in 1837 he noted a belt of large oaks (including one with a trunk 27 feet in circumference at 3 feet above the ground) "about three hundred yards in depth" (Belcher 1843). John Work (Mahoney 1945), in 1832, probably referring to French Camp Creek, a Sierra stream that flows to the delta, wrote: "the plain is overflowed and we had to encamp at the skirt of the woods about two miles from the river." Derby's report of 1849 (Farquhar 1932) noted a two-mile-wide belt of woods on both sides of the lower Feather River. The map accompanying this report shows forest bordering all the major and minor streams in the lower Sacramento River system. Thus, riparian forest seems to have bordered the entire mapped portion of the river system from the vicinity of Clarksburg in the south to Glenn in the north. These riparian forests are shown as being fairly uniform in width, about four to five miles. Derby's map also shows riparian forests along the tributary streams almost equal in width to those of the main stream, and flanking the tributaries to the edge of the valley. On the Derby map Cache and Putah creeks have forests about three miles wide, the American and Feather rivers about four miles wide (which checks with a section of his report), and Butte Creek and Yuba and Bear rivers each have levee forests about two miles wide. A note of caution should be inserted here. Derby, although a topographical engineer, performed only a reconnaissance type of survey of the valley. This being so, together with the undoubted fact that the tree symbols are intended to be approximate rather than precise, his map should not be invested with undeserved (and unintended) accuracy. However, even with these limitations the Derby map does suggest riparian forest of substantial width and continuity, and in 1849 these were, of course, still virtually in their pristine condition.

It is highly improbable that the forest belt was of uniform width along both banks of the streams. Indeed, historical accounts clearly indicate the irregular occurrence of the trees. Belcher (1837) refers to the trees as being "disposed in clumps." Derby also speaks of "clusters of beautiful trees-oaks, sycamore and ash" on the banks of the Yuba River to differentiate the forests there from those of the Sacramento and Feather rivers, which were "thickly wooded." Elsewhere he speaks of riparian forests along the Feather River "dotted" for two or three miles back from the river.

The Railroad Reports of a few years later (1855) speak of the riparian forest as being a "varying breadth, from a mile or more. . .to a meager border. Even more gener-ally, but clearly indicating the variation of width in the riparian forests, the Railroad Reports refer to the riparian forests as "of greater or less width." Moreover, the riparian forests varied not only in width but also in tree size and density, "the number and size of trees being apparently proportioned to the size of the stream and the quantity of moisture derived from it."

The preceding discussion shows that in their pristine condition the streams of the lower Sacramento River system were flanked by forests. The historical evidence suggests that these riparian forests had varied characteristics. They included trees of all sizes, from brush to very large valley oaks or sycamores, 75 to 100 feet high, growing closely spaced or scattered irregularly in groves. On the banks of the lower Sacramento, where the natural levees are widest, the riparian forests achieved their greatest width, four to five miles. On the lesser streams and in the delta, with smaller levees, the forests formed a narrower belt, generally about two miles wide but less in the delta. Dominant species in the riparian forest were valley oak (Quercus lobata), interior live oak (Quercus wislizenii), California sycamore (Platanus racemosa), Oregon ash (Fraxinus oregana), cottonwood (Populus fremontii), alder (Alnus rhombifolia), and several willows, (Salix gooddingii, S. exigua, S. Hindoiana, S. lasiandra, and S. laevigata).

Present Condition of the Riparian Forests

Although the Sacramento Valley riparian forests were an early casualty of the white man, their destruction, far-reaching as it was, was not complete. Today, parts of both banks of the Sacramento and its tributaries are bordered by many shrunken remnants of the once extensive riparian woodland. The numerous traces that remain corroborate the historical evidence examined by the author. The same tree species mentioned in the historical records - mainly valley oaks, cottonwoods, willows, sycamores, and ash - still grow on the river banks, natural levees, and channel ridges. Typically, cottonwoods and willows predominate on the immediate stream banks, whereas valley oaks are spread irregularly over the natural levees farther away from the river.

Instead of a strip measurable in miles, the forested zones along the Sacramento Valley streams are now often only yards deep, and discontinuous at that. Generally, the remaining fragments (not necessarily virgin stands, of course) form a belt less than 100 yards wide and are largely confined to bank slopes of streams and sloughs, abandoned meanders, and on the river side of artificial levees.

Examination of the Sacramento River levees reveals hundreds of larger relict stands of riparian forest. Some cover only a few acres; others several hundred. Most prominant are fully mature specimens of valley oaks in the "weeping" stage of development described by Jepson (1893) as indicating an age between 125 and 300 years. Such trees occur mostly on natural levee or channel ridge sites and are frequently around older settlements, presumably preserved for shade and ornament. Even small house lots may contain two or more oaks that predate the Anglo-American settlement period, presumably relics of a more extensive stand. Some tracts of uncleared land near the Sacramento River (including two in Yolo County between Knights Landing and Elkhorn Ferry) are still so thickly studded with trees, including many valley oaks in the "weeping" stage, that they form the definite, if open, forest described by early visitors to the region.

Near Woodson Bridge, Tehama County, another expanse of apparently virgin riparian forest can be seen. It is still subject to almost annual overflow and is composed mainly of mature valley oaks, forming an open woodland that extends discontinuously for about a mile from the river's edge. Some splendid mature specimens of valley oak remaining from the Cache Creek riparian forest can be seen in the older residential sections of Woodland in Yolo County, which is named for the fine oak forest in which the settlement was established in 1855. Again, in and around Davis, also in Yolo County, there are many large relict oaks of the Putah Creek forests.

In view of the general lack of trees in the Sacramento Valley, the riparian forests must have served as a source of fuel, construction, and other types of wood for a wide area. There was doubtless little incentive to conserve the riparian forests, since few of the tree species have much value as lumber. Typically the riparian forest species are fit only for low economic uses. For example, the numerous members of the genus Salix (willow) generally yield soft, light and brittle wood of poor form for saw timber. Rather similar is the cottonwood, which is soft, brittle, not durable, and especially liable to cracking. The largest, and probably most numerous, riparian tree, the valley oak, is "very brittle, firm, often cross-grained and difficult to split or work. On account of its poor timber form the trees are rarely if ever cut for anything but fuel, for which, however, they are much used." (Sudworth, 1908)

The clearing of the riparian forest for fuel and construction also served another end: it made available for agricultural use some of the most fertile and easily managed land in the valley. In its pristine, or nearly pristine, condition much of the valley was more or less unusable for agriculture because of waterlogging and inundations. The original limitations of many valley areas have been partially overcome in recent decades with improved drainage, irrigation, and other technical advances. However, initially these limitations were such as to discourage permanent settlement and agriculture on much of the valley floor with the exception of the natural levee lands. There both settlement and cultivation were concentrated; utilization of the remainder of the valley was uncertain and irregular, with much attention paid to livestock raising. The general superiority of the levee lands still holds. The most profitable form of land use in the valley, orchards, shows a very marked concentration on levee soils, a final confirmation of their inherent suitability for tree growth.

Perhaps because the riparian forests were largely effaced during the first two or three decades of Anglo-American occupance their existence is largely overlooked by modern students of the Sacramento Valley. But this neglected element in the landscape is by no means of negligible importance. The riparian trees served to reinforce the river banks and provide greater stability to the stream channels. They also acted as windbreaks, reducing evaporation, transpiration, and wind damage. In addition, the riparian forests provided a haven for the wildlife of the valley, furnishing cover and food sources for land and arboreal animals. Even more important was the fact that acorns, mainly from Quercus lobata, were a staple foodstuff of the Indian population. Furthermore, the forests furnished an important source of wood in an area otherwise poorly supplied.

The mere existence of the riparian forests, however, inevitably spelled their doom. The conditions, characteristic of natural levee sites, that permitted their development - comparative freedom from flood and waterlogging, high soil fertility, and favorable soil moisture - eventually led to their destruction, for the existence of the forest was incompatible with the modes of land use initiated by the Anglo-Americans. Today, only a few traces of the formerly extensive riparian forests remain, and the Sacramento Valley exhibits a striking lack of trees.

Literature Cited

Belcher, Captain Sir Edward, R.N., "Narrative of a Voyage Round the World Performed in Her Majesty's Ship Sulphur During the Years 1836-1842." Vol. I (London: Henry Colburn, 1843), p. 130.

Cook, S. F. 1955. "The Epidemic of 1830 - 1833 in California and Oregon." University of California Publications in American Archaeology and Ethnology, Vol. 43, No. 3.

Farquhar, Francis P. (ed.). 1932. "The Topographical Reports of Lieutenant George H. Derby," California Historical Society Quarterly, Vol. II, p. 115.

Jepson, Willis L. 1893. "The Riparian Botany of the Lower Sacramento." Erythea Vol. 1 p. 242.

Mahoney, Alice B. (ed.) 1945. "Fur Brigade to the Bonaventura, John Work's California Expedition 1832-33 for the Hudson's Bay Company (San Francisco: California Historical Society), p. 18.

Sudworth, George B. 1908. Forest Trees of the Pacific Slope (Washington, D.C.: Department of Agriculture), pp. 212-278.

Reproduced by permission from the ANNALS of the Association of American Geographers, Volume 51, 1961, K. Thompson.

Chapter 5

THE FLUVIAL SYSTEM: SELECTED OBSERVATIONS

E. A. Keller
Environmental Studies and
Department of Geological Sciences
University of California Santa Barbara
Santa Barbara, California 93106

Introduction

Human use and interest in the riverine environment extends back to earliest recorded history. We have used the river system as an avenue for transportation and communication, a water supply, a waste disposal site, and a source of power. Massive dams and channel works to dissipate the disastrous effects of floods and droughts have been constructed, and even though we can sometimes control a river we still know little about the processes which form and maintain the natural fluvial system. Only recently have we realized that rivers are natural resources that must be conserved and properly managed if we are to continue a meaningful existence.

The natural riverine environment is basically an open system composed of three inter-related parts or phases: 1) the fluid or water; 2) the channel and floodplain; and 3) the network of channels comprising the drainage basin. As the system evolves and changes all three parts mutually adjust and each exerts a partial control on the others. The communication between the various parts in the riverine system includes a multitude of interactions which tend to maintain a delicate balance or equilibrium within the system. This equilibrium that develops in most streams is the quasi-equilibrium discussed by Leopold and Maddock (1953) or dynamic equilibrium described by Hack (1960).

The primary purpose of this paper is to present fundamental concepts necessary for understanding the fluvial system. These will include: 1) recognizing that the stream channels and adjacent floodplain comprise an erosional, transportational and depositional environment in which form and process evolve in harmony; 2) recognizing that significant changes in the fluvial system often occur when a geomorphic or hydraulic threshold is exceeded; and 3) recognizing that human interference with the fluvial system generally reduces the channel, floodplain and hydrologic variability.

Fundamental Concepts

Stream Channel-Floodplain Environment

The stream channel and adjacent floodplain comprise a unique environment in which the floodplain, channel and bedforms evolve in harmony with natural fluvial processes that erode, transport, sort and deposit alluvial bed and bank materials.

Floodplain: Processes significant in the formation of a floodplain include: 1) overbank flow and resultant deposition; and 2) lateral migration of the stream channel. Of the above two processes lateral migration of the stream channel is probably most significant in the formation of floodplains.

The natural stream channel generally has sufficient discharge to emerge from its banks and flood areas adjacent to its banks on the average of once every year or two. It is this natural process of overbank flow which slowly but relentlessly builds floodplain features such as natural levees along the stream channels. The overbank flows also supply water to adjacent lowlands on the floodplain which serve as a storage site for excess runoff, much of which may enter the groundwater system. A main philosophical concession that must be recognized by more communities which compete with the riverine environment is that overbank flow (flooding) is a natural process rather than a natural hazard and that, if we are to maintain the integrity of the riverine system, we must consider the channel and floodplain as a complementary system.

The most significant process that builds floodplains is lateral migration of stream channels. Evidence for this is based on the fact that a meandering stream channel migrates continuously across the floodplain with time and, in doing so, processes of erosion and deposition take place which continuously create and modify the floodplain. Evidence such as meander scrolls, oxbow lakes, and meander scars on many floodplains attest to the fact that the position of the meandering stream changes constantly. The actual process of lateral migration occurs as the stream erodes its channel on the outside of a meander bend and deposits on the inside. Thus, the meander bend moves laterally and down valley while maintaining a constant channel geometry. The actual erosion of the channel generally takes place following rather than during a flood event. In many streams the flood discharge drops rather quickly and this leaves considerable unsupported water in the stream banks. Thus an adverse water pressure in the stream banks facilitates slumping on the outside of meander bends, and material that falls into the channel is removed by subsequent flows.

Channel Pattern: The pattern of an individual stream channel in the planemetric view is termed the channel pattern. Natural streams fall into two major patterns: 1) braided channels characterized by the existence of numerous islands and bars which continuously divide and reunite the channel; and 2) channels that are not braided. Because stream channels that remain straight for any significant length of channel are relatively rare, channels that do not braid may be designated as sinuous. However, even sinuous streams often have short straight reaches. These straight reaches may be produced by several processes including: 1) lateral migration of adjacent meander bends which tend to increase sinuosity and thus locally decrease channel slope; and 2) meander cutoffs which reduce the sinuosity of a channel and thus locally increase the channel slope (Keller and Melhorn, 1973). Furthermore, it is important to recognize that although a straight channel has straight streambanks it implies neither a uniform stream bed nor a straight thalweg (Leopold and Wolman, 1957). Sinuosity, as used in this paper, refers to the ratio of channel length to valley length and the thalweg is the line delineating the deepest portion of the stream channel in the downstream direction.

Meandering channels may be considered as a special type of sinuous channel characterized by a series of relatively symmetrical curves with sinuosity greater than 1.5. However, the precise definition of symmetry necessary to designate a channel as meandering has not been established, and therefore it seems advantageous to use the term "meandering" rather loosely. Thus, as suggested by Leopold, et al. (1964), we will define a meandering channel as a sinuous stream with sinuosity greater than 1.5 irrespective of the symmetry of the channel bends.

Fluvial Hydrology: In many alluvial stream channels the characteristic forms are produced at relatively high channel forming flows and are only modified at low flow. Therefore, conventional hydrology applies at low flow (with little sediment transport) when the channel is essentially a rigid container for the liquid or fluid phase. However, at high flow when appreciable sediment is being transported and sorted by the stream, conventional hydrology is no longer applicable because many of the variables are not unique (Maddock, 1969, and Leliavsky, 1966). Thus, we need to distinguish fluvial hydrology from hydraulics in general as a necessary step in understanding the natural riverine environment.

Important principles of fluvial hydrology include: 1) in no part of the natural alluvial stream channel are contiguous stream lines parallel to one another or to the bank of the stream; 2) the greater decurvature of the horizontal projections of the stream projectories, the deeper channel scours below them; and 3) at high channel-forming flow, scour in stream channels is associated with convergence of stream flow and deposition is correlated with divergence of flow (Leliavsky, 1966). This last principle of fluvial hydrology is known as the convergence-divergence criterion (Figure 1). In general during channel-forming flood events convergence of flow is typical of channel deeps (pools), and divergence of flow is typical of shallows (riffles).

Verification of the convergence-divergence criteria in natural streams comes from its application to river training (Leliavsky, 1966, and Keller, 1976). Unfortunately, the full potential and significance of the criterion has not been realized in studying natural streams. In part this results from the fact that with little or no sediment transport, at low flow, when most rivers are studied, the observed convergence and divergence of flow may not be operating in pools and riffles. During this stage normal hydrology is applicable. However, at high flow with moving sediment, fluvial hydrology applies and convergence of flow in deeps is probably responsible for the observed scour. The conclusion is that major forms of scour and deposition in stream channels are apparently produced at relatively high channelforming flows and are modified only at low flow. Therefore, characteristic forms surveyed at low flow in stream channels may depict relic forms.

<u>Bed Forms</u>: The most obvious forms in a stream channel, excluding the fluvial phase, are bed forms. A bed form may be defined as any irregularity produced on the bed of an alluvial stream channel by the interaction of the flowing water and moving sediment (Simons and Richardson, 1966). In a meandering river with appreciable gravel in the stream bed, there are two main types or groups of bed forms: 1) pools, riffles, and point bars which give the stream channel its basic morphology; and 2) ripples, dunes, and antidunes which are mostly controlled by the hydrologic phase and are not generally a significant part of the basic channel morphology (Keller and Melhorn, 1973). Pools, riffles and point bars are best developed in gravel-bed alluvial stream channels with a small sand fraction in the stream bed. Stream channels with a significant amount of sand in the channel produce migrating ripples and dunes at low flow which may be superposed on and partly mask the more stable bed forms.

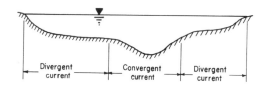

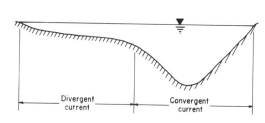

After Leliavsky, 1966

Figure 1. Convergence-divergence criterion for common river cross sections.

Pools, riffles, and point bars may be defined by their basic morphology (Keller, 1971a).

<u>Pool</u>: a topographically low area produced by scour which generally contains relatively fine bed material. Pools are generally associated with a point bar.

<u>Riffle</u>: a topographic high area produced by the accumulation of relatively coarse-grained bed material. Ideally, the inflection point of the thalweg is located on the riffle between successive pools. At the inflection, the cross profile is generally symmetrical.

<u>Point bar</u>: an accumulation of relatively coarse-grained bed material on the concave side of the thalweg adjacent to the pool. The pool and point bar together produce an asymmetric cross profile. Basic morphology of a gravel-bed, sinuous (meandering) stream channel is generalized on Figure 2.

Pools and riffles at low flow are recognized by regularly repeating deeps and shallows (Figure 3). Pools are areas of relatively deep, slow moving water with a low water-surface gradient, whereas riffles are recognized by relatively shallow, fast-flowing water with a steeper water-surface gradient. As discharge increases, the difference in water-surface gradient between pools and riffles gradually disappears until about at bank-full stage when the water-surface gradient is constant and the pool-riffle sequence is said to be "drowned out" (Leopold, et al., 1964). However, it is significant to recognize that the basic undulating bottom topography of the pool-riffle sequence does not disappear at high flow.

Pools and riffles are significant in the riverine environment because: 1) pools and riffles at low flow conditions provide a variety of flow conditions varying from slow deep water to fast shallow water that is necessary for feeding, breeding, and cover for riverine life forms; 2) pools and riffles at high flow provide shelter areas with protection from excessive water velocity for fish and other life forms; 3) pools and riffles maintain a natural sorting of bed-load materials such that coarse material is found on riffles and finer material is found in pools, providing a good environment for bottom-dwelling organisms; 4) pools and riffles along stream channels facilitate a diversity of stream bank vegetation necessary in providing cover and food for riparian organisms; and

5) pools and riffles provide a diversity of sensual stimuli and physical contrasts such as shaded versus sunlit water, shallow bubbling water on riffles versus slower water in pools, and light green or brown leaves of trees and brush along banks versus deep green reeds and rushes near the stream (Eiserman, et al., 1975, and Keller, 1976).

Many alluvial stream channels are characterized by regularly spaced pools and riffles. In these channels, pools and riffles remain in approximately the same location over a series of flow and, therefore, these channels may be considered as being morphologically stable. Measurement of 102 pool-riffle sequences in California and Indiana streams (Figure 4) strongly confirms the conclusion of Leopold, et al. (1964) that pools and riffles are spaced on the average of five to seven times the channel width. The channel width is here defined as the width of the active stream bed delineated by the existence of bed material. Channel width is generally measured at riffle sites where the cross channel profile is symmetrical. Channel width as defined above, is close to the bank-full width. The similarity of spacing of pools and riffles in both straight and meandering channels suggests that the development of pools and riffles is one of the primary fluvial processes occurring in streams (Keller, 1972). Furthermore, the tendency to develop pools and riffles is apparently associated with some type of wave phenomena that is significant in the meandering process (Leopold, et al., (1964).

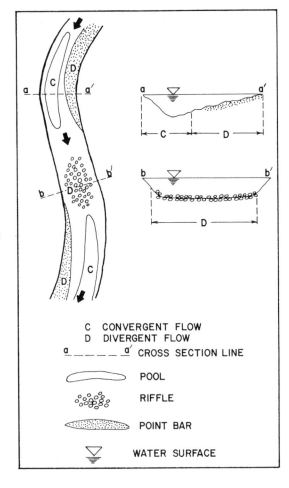

Figure 2. Basic channel morphology for the meandering river with a gravel bed.

Figure 3. Well-developed pools and riffles in Sims Creek near Blowing Rock, North Carolina.

Geomorphic and Hydraulic Thresholds:

Many changes that take place in the fluvial system are in an episodical or pulsating manner that tend to be controlled by geomorphic or hydraulic thresholds. Two types of fluvial thresholds can be identified: 1) those associated with negative feedback; and 2) those associated with positive feedback.

Probably the best known of the hydraulic thresholds in fluvial systems is the threshold defined as the velocity necessary to initiate particle motion along the bed of a stream. Essentially, this threshold is characterized by positive feedback, since initiation of movement of a particle facilitates the movement of other particles. Another well-known fluvial hydraulic threshold occurs when the Froude number exceeds one. Bed forms associated with tranquil flow and Froude numbers less than one differ in many ways from those bed forms associated with Froude numbers greater than one. The changes initiated by a triggering mechanism delineated mathematically by a critical value (Froude number) when the ratio of the inertia force to the gravity force of the flowing water exceeds one.

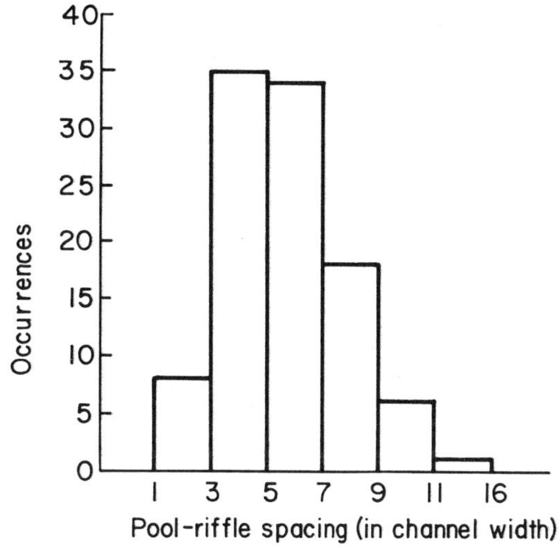

Figure 4. Frequency of pool-riffle spacing for four alluvial streams (102 pool-riffle sequences).

Alluvial streams with well-developed pools and riffles produce another type of hydraulic threshold. Pools at low flow are characterized by deep slow moving water and riffles by faster shallow water. However, at high flow the situation is reversed and pools are areas with the highest velocity of water (Keller, 1971b, and Andrews, 1976). This concept is consistent with Leopold's and others' conclusions that with increased discharge the water slope over pools increases faster than that of adjacent riffles. This process has been termed "Hypothesis of Velocity Reversal." That is, as discharge increases over a pool-riffle sequence, the initial velocity in pools is less than that of adjacent riffles. However, eventually a threshold is crossed and at that point the velocity in the pool may exceed that in the riffle. The implication of the reversal is very important in scouring out pools and depositing coarse bed materials on riffles during high channel-forming flows. That is, pools scour at high flow, fill at

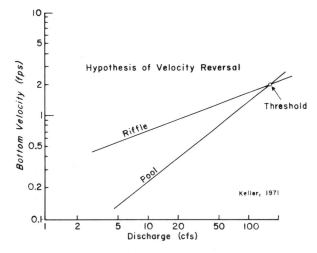

Figure 5. Velocity reversal (threshold) concept: data for Dry Creek near Winters, California.

43

low flow, whereas riffles fill at high flow and scour at low flow. This scour and fill pattern is controlled by a hydraulic threshold with a negative feedback mechanism that allows pools and riffles to persist over a period of flows (Figure 5).

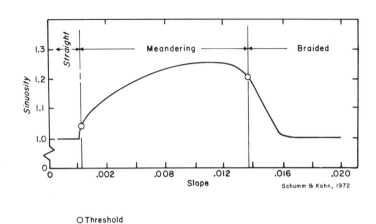

Figure 6. Threshold concept: sinuosity (channel pattern) and channel slope.

Geomorphic thresholds also exist which tend to partially control the form of a stream channel. Perhaps the best known example includes threshold values of channel slope which tend to control channel pattern (Figure 6) as defined by Schumm and Kahn (1971); however, it should be recognized that these thresholds were developed in a flume and as such may not be directly applicable to natural stream channels. The main point, however, is that the changes in channel pattern rather than being continuous tend to occur as threshold values are exceeded. Following the change, the feedback mechanism seems to be in a negative direction--that is, to maintain a quasi- or dynamic equilibrium within the stream. In other words, a meandering alluvial stream will increase in sinuosity and develop meander cutoffs periodically which locally increase its slope and reduce sinuosity. This may initiate a local change in channel pattern, the result of which is to help maintain the quasi- or dynamic equilibrium of the river.

Another type of geomorphic threshold in the fluvial system involves the process of lateral migration of meander bends in relatively cohesive bank materials. It has been noted that most of the lateral migration may be due to bank caving or slumping. However, the process of bank caving does not occur at maximum rates during flood events. This results because the water in the stream helps to support the channel-bank materials. However, following rapid draw-down of water levels following the flood event, stream banks on the outside of bends may fail quite suddenly. This occurs as a threshold value in the shear strength of the bank material is exceeded. This type of threshold is probably a feedback mechanism in the adjustment of channel slope.

The conclusion reached from our discussion of geomorphic and hydraulic thresholds is that many changes that take place in alluvial stream channels do so in response to stream processes in which a threshold is exceeded. The changes that take place may be in a direction towards a positive feedback or disequilibrium or negative feedback which tends to maintain the fluvial system.

Human Interference with the Fluvial System:

Human use and interest in the fluvial environment has historically included significant drainage modification. This modification--whether termed channelization, channel works, or channel improvement--generally is controversial because of its potential adverse effects on the biological communities in the riverine environment. The loss of fish and wildlife habitat due to channel modification generally leads to simplification with less variation in the biological communities of the fluvial environment. Figure 7 contrasts some of the differences between natural channels and man-made channels that lead to environmental deterioration (Corning, 1975).

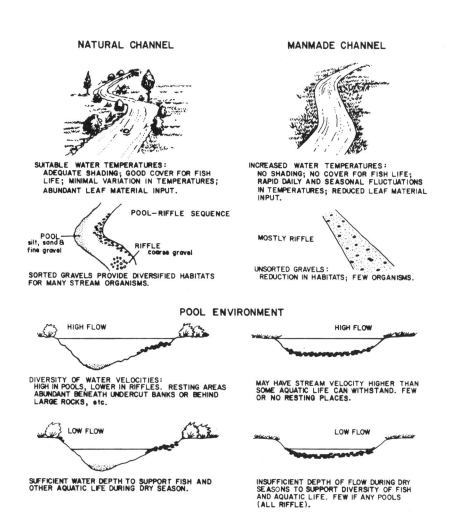

NATURAL CHANNEL

SUITABLE WATER TEMPERATURES:
ADEQUATE SHADING; GOOD COVER FOR FISH
LIFE; MINIMAL VARIATION IN TEMPERATURES;
ABUNDANT LEAF MATERIAL INPUT.

POOL-RIFFLE SEQUENCE

POOL
silt, sand &
fine gravel

RIFFLE
coarse gravel

SORTED GRAVELS PROVIDE DIVERSIFIED HABITATS
FOR MANY STREAM ORGANISMS.

MANMADE CHANNEL

INCREASED WATER TEMPERATURES:
NO SHADING; NO COVER FOR FISH LIFE;
RAPID DAILY AND SEASONAL FLUCTUATIONS
IN TEMPERATURES; REDUCED LEAF MATERIAL
INPUT.

MOSTLY RIFFLE

UNSORTED GRAVELS:
REDUCTION IN HABITATS; FEW ORGANISMS.

POOL ENVIRONMENT

HIGH FLOW

DIVERSITY OF WATER VELOCITIES:
HIGH IN POOLS, LOWER IN RIFFLES. RESTING AREAS
ABUNDANT BENEATH UNDERCUT BANKS OR BEHIND
LARGE ROCKS, etc.

LOW FLOW

SUFFICIENT WATER DEPTH TO SUPPORT FISH AND
OTHER AQUATIC LIFE DURING DRY SEASON.

HIGH FLOW

MAY HAVE STREAM VELOCITY HIGHER THAN
SOME AQUATIC LIFE CAN WITHSTAND. FEW
OR NO RESTING PLACES.

LOW FLOW

INSUFFICIENT DEPTH OF FLOW DURING DRY
SEASONS TO SUPPORT DIVERSITY OF FISH
AND AQUATIC LIFE. FEW IF ANY POOLS
(ALL RIFFLE).

Figure 7. Comparison of a natural and man-made channel (modified after Corning, 1975).

The reduced variability of the biological community in response to channel modification is directly attributed to the loss of variability in the physical environment. That is, stream channel modification tends to reduce the diversity of flow conditions, the diversity of bed-material distribution, and the diversity of bed forms. If environmental deterioration caused by stream channel modification is to be minimized then new design criteria must be developed such that the stream's natural tendency to converge and diverge flow and sort the bed material is maintained (Keller, 1975). That is, we must apply environmental determinism or "designing with nature" to our channel works if we are to maintain a quality fluvial environment.

Conclusions and Recommendations

The natural fluvial environment is an open system in which the channel-floodplain form and processes evolve in harmony. Significant changes in the fluvial system often occur when a geomorphic or hydraulic threshold is exceeded. These changes are partly responsible for maintaining the quasi- or dynamic equilibrium state of the stream system. Human use and interest in the fluvial environment has led to human interference with the fluvial system. This interference generally reduces the channel, floodplain and hydraulic variability and thus the biologic variability which depends on the physical environment.

The behavior of natural streams is not completely understood. Particularly important is the need to know more about relationships between erosion, deposition, and sediment concentration, as well as the effect of organic debris on stream channel morphology. In addition, if we are going to understand more about relationships between the biology of stream channels and the geomorphology then we must begin to study complex interactions between the two. That is, we must learn more about processes which produce channel

45

morphology necessary for biological productivity and thresholds that control the maintenance and development of the physical and biological environment.

Literature Cited

Andrews, E. D., 1976, River Channel scour and fill; (abstract), Geol. Soc. Amer. Abstracts with Programs, V. 8, No. 6, p. 755.

Eiserman, F., Dern, G., and Doyle, J., 1975, Cold water stream handbook for Wyoming, Soil Conservation Service and Wyoming Game and Fish Dept., 38P.

Hack, J. T., 1960, Interpretation of erosional topography in humid temperate regions, Amer. Jour. Sci., V. 258, pp. 80-97.

Keller, E. A., 1971a, Pools, riffles and meanders: discussion, Geol. Soc. Amer., V. 82, pp. 279-280.

Keller, E. A., 1971b, Areal sorting of bed-load material: the hypothesis of velocity reversal, Geol. Soc. Amer. Bull., V. 82, pp. 753-756.

Keller, E. A., 1972, Development of alluvial stream channels: a five stage model, Geol. Soc. Amer. Bull., V. 83, pp. 1531-1536.

Keller, E. A., 1976, Channelization: environmental, geomorphic and engineering aspects, in D. R. Coates (ed.), Geomorphology and Engineering, Dowden, Hutchinson and Ross, Inc., Stroudsburg, Penn., pp. 115-140.

Keller, E. A., and Melhorn, W. N., 1973, Bedforms and Fluvial processes in alluvial stream channels: selected observations, in M. Murisawa (ed.) Fluvial Geomorphology, State Univ. N.Y., Binghamton, N.Y., pp. 253-284.

Leliavsky, S., 1966, An introduction to fluvial hydraulics, Dover, N. Y., pp. 102-106.

Leopold, L. B., and Maddock, T., Jr., 1953, The hydraulic geometry of stream channels and some physiographic implications, U. S. Geol. Survey Proj. Paper 252, 57P.

Leopold, L. B., and Wolman, M. G., 1957, River Channel Patterns: braided, meandering and straight, U. S. Geol. Survey Prof. Paper 282B, pp. 53-59.

Leopold, L. B., Wolman, M. G., and Miller, J. P., 1964, Fluvial processes in geomorphology, W. H. Freeman and Co., San Francisco, 522P.

Maddock, T., Jr., 1969, The behavior of straight open channels with movable beds, U. S. Geol. Survey Prof. Paper 622A, 70P.

Schumm, S. A. and H. R. Kahn, 1971, Model study of river patterns, Geol. Soc. Amer. Abstr. Vol. 3 No. 7, p. 697.

Simons, D. B., and Richardson, E. V., 1966, Resistance to flow in alluvial channels, U. S. Geol. Survey Prof. Paper 422J, 61P.

Chapter 6

RIPARIAN VEGETATION AND FLORA OF THE SACRAMENTO VALLEY

Susan G. Conard, Rod L. MacDonald, Robert F. Holland
Departments of Botany and Agronomy and Range Science
University of California, Davis 95616

Introduction

Prior to the extensive human alteration of California's vegetation over the past two centuries, the major streams of the Sacramento Valley were bordered by riparian forests and woodlands, which occurred in bands up to 10 miles wide on the coarse (sand to silt) alluvium of natural levees and river terrace deposits. (Thompson, 1961). Woodcutting by early settlers and, more recently, agricultural encroachment, have considerably reduced the extent of these forests. From 1952 to 1972 the acreage of native riparian vegetation on alluvial soils along the Sacramento River between Redding and Colusa decreased 34% from 27,720 ac. (11,267 ha) to 18,360 ac. (7,435 ha). This loss of riparian habitat was primarily due to conversion to orchards and annual row crops (McGill, 1975). By 1972, croplands covered approximately 66% of the terrace lands which once supported riparian forest. The use of flood control levees rather than soil types to delineate mapping boundaries for this study has undoubtedly resulted in an underestimate of the percentage of riparian forests converted to agriculture. Conversion has been close to 100% beyond flood control levees. Few stands large enough for quantitative vegetation study remain.

Riparian forests are the only forests in California dominated by mesophytic, broadleaf, winter-deciduous taxa. The large number of genera in common with the deciduous forests of the Eastern United States and Asia, including Acer, Populus, Platanus, Fraxinus, Quercus, Alnus, and Juglans, (Robichaux, 1977) show the close floristic affinities to these regions. Many species are restricted to riparian habitats within California. Riparian and marsh settings in the Sacramento Valley are unique in having water available for plant growth during the warm, rainless Mediterranean summer. This also eliminates drought hardening, and hence resistance to freezing temperatures, for these species. These conditions favor the development of broad-leaved, deciduous vegetation (Stebbins, 1971).

Research on Sacramento Valley riparian vegetation has primarily concerned land-use patterns (McGill, 1975; Brumley, 1976) or distribution and ecology of birds and mammals in riparian habitat (Stone, 1976; Michny, Boos, and Wernette, 1975; Brumley, 1976; Stroud, 1977). These studies frequently include partial floristic lists or brief vegetation descriptions. Michny, et al. (1975) provide quantitative vegetation data at nine study sites. Most of these stands were apparently less than 15 m. in width and several were highly disturbed.

Objectives

The objectives of this study were: 1) to obtain preliminary floristic and vegetation data on several major riparian vegetation types, and 2) to use such data to, a) delineate important vegetation units, b) describe structure of mature stands of riparian forest, and c) describe major seral and topographic relationships within the riparian vegetation. Field work was in conjunction with a palynological study conducted by Jim West (1977). To minimize contamination of pollen samples and effects of human activities peripheral to the stands, site selection was limited to stands of continuous vegetation exceeding forty acres (16 ha). All sites showed minimal evidence of human disturbance. Although the oak woodland areas are used periodically for pasture lands, grazing pressure appears slight to moderate. Eight study sites were selected on the basis of areal and ground reconnaissance and are identified in Figure 1. Sites were selected to represent high terrace valley oak woodlands (Cosumnes, Sac R. 1); low terrace riparian forest (Glenn East, Glenn West, Snodgrass Slough); and seral vegetation types (Sac R.2, Glenn West, Snodgrass Slough).

Vegetation Composition

At each study site vegetation was sampled along topographic physiographic gradients extending from the river up to the highest terrace (Figures 2 & 3). Homogeneous stands of vegetation were subjectively located and sampled by the "Braun-Blanquet" or "releve" method

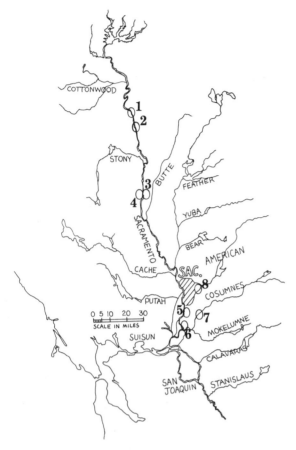

Figure 1. The location of riparian vegetation study sites.

1. Sacramento River 1

2. Sacramento River 2

3. Glenn East

4. Glenn West

5. South Stone Lake

6. Delta Meadows-Snodgrass Slough

7. Cosumnes River

8. American River

(Mueller-Dombois and Ellenberg, 1974, Chapter 5). Each releve consists of a list of all species encountered in the stand of vegetation, annotated with visual estimates of cover (percent of ground covered by the projected canopy of each species). Cover values were later transformed to the following abundance scale (following Braun Blanquet - see p. 147 in Mueller-Dombois and Ellenberg, 1974): R = single individual, + = <1% cover, 1 = 1-5%, 2 = 6-25%, 3 = 26-50%, 4 = 51-75%, 5 = 76-100%.

The information obtained in the stand survey was encoded on computer data cards. The data set was processed in a burroughs 6700 using an association analysis program developed by Ceska and Roemer (1971) as adapted by Drs. David Randall and Dean Taylor at Davis. This process identifies associational relationships between the species according to pre-established phytosociological criteria (see Table 1).

A species belongs to an associational group if it occurs in 50% of the stands of that group and does not occur in more than 20% of the remaining stands. A stand belongs to an associational group when 50% of the species in that group are present in the stand.

Vegetation Structure

Basal areas and densities for trees greater than 10 cm DBH (diameter at 140 cm above the ground) were determined by the point-centered quarter method (Cottam and Curtis, 1956). Points were located 30 m. apart along compass transects approximately paralleling natural river channels. Diameter (to nearest 5 cm.) and species were recorded for each tree.

Results and Discussion

The positions of the major riparian plant communities in the floodplain, along with representative species, are illustrated diagrammatically in Figures 2 and 3. The association table (Table 1) presents data from stand surveys grouped according to similarities in species composition. The major riparian vegetation types which emerge are 1) Valley oak woodland, 2) Riparian forest dominated by cottonwood, 3) Gravel bar thickets, 4) Open flood plain communities, and 5) Hydric communities.

Valley Oak Woodland

The valley oak phase of the riparian forest is typical on high terrace deposits and above cut banks along the outside of meanders. These forests are dominated almost exclusively by valley oak (Quercus lobata). Common associates include sycamore (Platanus racemosa), willows (Salix spp.), box elder (Acer negundo), Oregon ash (Fraxinus latifolia), and black walnut (Juglans hindsii). Canopy height is 15-20 m and canopy cover ranges from 30-60%. A typical valley oak woodland sampled at the Cosumnes site (Table 2) had a density of 124.5 trees/ha and basal area of 18.35 m^2/ha. The relative density of Q. lobata in this stand is .73 and its relative basal area is .81, indicating strong dominance by Q. lobata at this site.

Table 2. Density and basal areas of trees greater than 1 dm. DBH. Cosumnes River site. n = 84

OAK WOODLAND -- COSUMNES RIVER

	Density (trees/ha)	Relative Density	Basal Area (m^2/ha)	Relative B.A.	BA/tree (dm^2)
Quercus lobata		.73		.81	16.5
Fraxinus latifolia		.13		.10	11.1
Salix lasiolepis		.05		.05	14.4
Acer negundo		.04		.03	13.8
Cephalanthus occidentalis		.04		<.001	2.1
Prunus armeniaca		.01		<.001	1.8
Juglans hindsii		.01		<.001	1.1
TOTAL	124.5	1.0	18.35	1.0	14.7

Valley oak woodlands are characteristically heterogeneous with areas of high density, smaller trees interspersed with more open areas of larger trees. Openings contain typical grassland species of genera such as Avena, Lolium, Hordeum, and Elymus (Table 1). Where tree cover is higher, the understory is characterized by poison hemlock (Conium maculatum), poison oak (Toxicodendron diversiloba), ripgut brome (Bromus diandrus), soap plant (Chlorogalum pomeridianum), several species of Carex, and Erigeron sp.

Riparian Forest

Cottonwood (Populus fremontii) dominates the riparian forest of lower terrace deposits and stabilized gravel bars along the Sacramento River. Common associates are similar to those in the valley oak woodland including willows (Salix lasiolepis, S. gooddingii, S. laevigata, S. lasiandra), Fraxinus latifolia, Acer negundo, Juglans hindsii, and on higher ground, Quercus lobata and Platanus racemosa (Table 1). Canopy height is approximately 30 m. in a mature riparian forest, with a canopy cover of 20-80%. Tree density (Table 3) in these forests is about 250 stems/ha - double that of valley oak woodland sampled. Basal area is about 40 m^2/ha. The relative basal area of Populus fremontii is .75, reflecting its high dominance in the vegetation. The low relative density (.33 - .44) of cottonwood in these stands reflects the large number of small subcanopy (10-12 m) trees (particularly Acer negundo, Fraxinus latifolia, and Salix spp). Understory species are mostly shrubs (Sambucus mexicana, Cephalanthus occidentalis, Rubus spp, Rosa californica). Lianas such as Rhus diversiloba and Vitis californica are a dominant feature, frequently providing 30-50% ground cover and festooning trees to heights of 20-30 m. Herbaceous vegetation is 1% cover except in openings where tall forbs such as Artemisia douglasiana, Urtica dioica, and various shade tolerant grasses may occur.

Table 1. The composition of selected riparian vegetation summarized in a species-by-stand matrix. The arrangement of the species and stands reflects association analysis using the Ceska-Roemer technique. CONS - Cosumnes River Site, Sacramento Co.; GLENN - sites near Princeton, Glenn County.

Species	GRASSLAND							VALLEY OAK WOODLAND											RIPARIAN FOREST						SERAL STANDS	
	CONS15	CONS13	CONS14	CONS5	CONS6	CONS16	CONS7	CONS9	CONS10	CONS12	CONS3	CONS19	CONS18	CONS11	CONS14	CONS8	CONS17	CONS5	GLENN17	GLENN13	GLENN12	GLENN22	GLENN8	GLENN6	GLENN4	GLENN1
Lolium multiflorum	3	+	1	3	1	3	2																			
Avena fatua	4	1	2	2	1	1	1	+	+																	
Hordeum leporinum	1	2	2	3	2	+	2																	+		
Hordeum geniculatum		2	2	2	2	+																				
Hordeum brachyantherum	1	+	+	R	R	3																				
Taraxicum officinale											1													+		
Elymus triticoides	1			1	1																					
Ranunculus californicus	1		+		+																					
Geranium dissectum	1				+																					
Convolvulus arvensis	1				1			2		2						3										
Carex 1	R																									
Cerastium viscosum					+																					
Vicia sativa											2															
Quercus lobata							1	3	4	4	3	3	2	3	3	3	5	4	3	3						1
Conium maculatum								2	2	1	3	1	+	1	1	1	1	+	+	1						
Rhus diversiloba								2	2	1	1	1	1	1	2	1	2		+	2						
Bromus diandrus					1		+	2	3	2	2		2	3												
Chlorogalum pomeridianum								1	2	1	1	1		1												
Erigeron								2	1	1	1															
Carex 2								2	1	3	3															
Carex 3								1	1	+	1	3														
Populus fremontii																	1	1	2	4	2	1	1	2	2	1
Vitis californica																	2	1	2	1	1	2		2	1	
Fraxinus latifolia												R					2	1	R	1	R	1			1	
Rubus leucodermis								+									4	1	4	2	3	2	2			
Rosa californica													1		1	2	1	1	1	+	2	1				
Acer negundo var. californicum																			2	2	2	1		1	2	1
Aristolochia californica																	+	+	1						+	
Clematis ligusticifolia																	+	+	2						+	

Table 1, continued.

Species	GRASSLAND						VALLEY OAK WOODLAND											RIPARIAN FOREST SERAL STANDS									
	CONSU15	CONSU34	CONSU45	CONSU16	CONSU67	CONSU79	CONSU90	CONSU22	CONSU13	CONSU19	CONSU18	CONSU14	CONSU18	CONSU17	CONSL75	CONST78	CONST7	GLENN51	GLENN17	GLENN21	GLENN32	GLENN33	GLENN22	GLENN81	GLENN64	GLENN41
Sambucus mexicana															2				1					R		
Juglans hindsii								2			2									1			3			1
Cephalanthus occidentalis																		2 / R		3						
Platanus racemosa																										
Baccharis pilularis																					3	3				
Rubus vitifolius									+		2							1								
Symphorocarpus rivularis							+		+		2 / 1							+ / +	+ / +				1			
Artemesia douglasiana		1		1			1	2	+	2	1							R	R				3 / 1			
Rumex crispus							2	2	2	+													3			
Glyceria sp.																+	5							2		
Salix hindsiana																										
Bidens frondosa																		+		1						
Sorgum halepense				+														+	+		3	3				
Epilobium sp.	1	1											+													
Raphanus sativa				1		2					+							2							+	
Solidago Sp.			3												2					2						
Brassica campestris								2			+				+											+
Prunus sp.			R			2												1	1							+
Polypogon monspeliensis									1 / +																	
Xanthium strumarium var. canadense																										
Bromus mollis		3																1								
Anaphalis margaritacea																							1			
Lolium perenne																										
Bromus sp.																										
Polygonum sp.			1																							
Rorippa islandica var. occidentalis																							1			
Medicago hispida																										
Plantago lanceolata															+											
Poa annua										2																
Coniza canadensis																				1					+	
Anagallis arvensis																										+
Brassica kaber																							1			
Festuca megalura																										
Nepeta cataria																										+

51

Figure 2. IDEALIZED TOPO-SEQUENCE OF RIPARIAN VEGETATION ALONG MAJOR RIVERS -- SACRAMENTO VALLEY, CA.

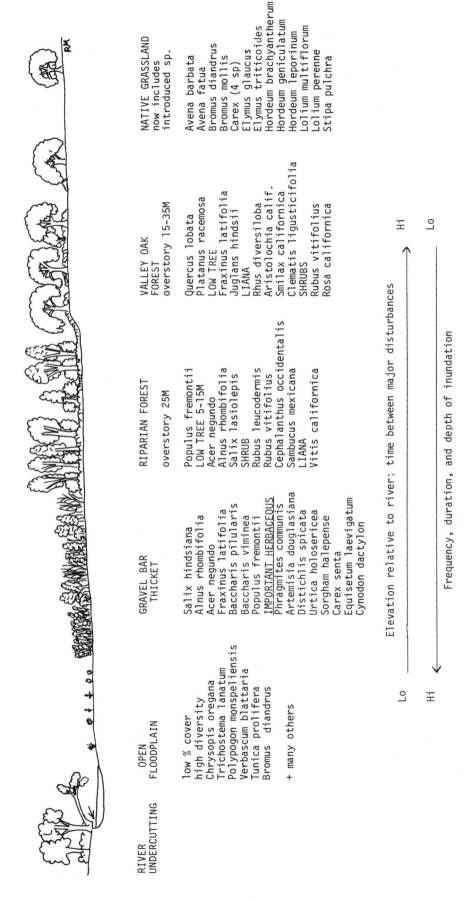

RIVER UNDERCUTTING	OPEN FLOODPLAIN	GRAVEL BAR THICKET	RIPARIAN FOREST overstory 25M	VALLEY OAK FOREST overstory 15-35M	NATIVE GRASSLAND now includes introduced sp.
	low % cover high diversity Chrysopis oregana Trichostema lanatum Polypogon monspeliensis Verbascum blattaria Tunica prolifera Bromus diandrus + many others	Salix hindsiana Alnus rhombifolia Acer negundo Fraxinus latifolia Baccharis pilularis Baccharis viminea Populus fremontii IMPORTANT HERBACEOUS Phragmites communis Artemisia douglasiana Distichlis spicata Urtica holosericea Sorghum halepense Carex senta Equisetum laevigatum Cynodon dactylon	Populus fremontii LOW TREE 5-15M Acer negundo Alnus rhombifolia Salix lasiolepis SHRUB Rubus leucodermis Rubus vitifolius Cephalanthus occidentalis Sambucus mexicana LIANA Vitis californica	Quercus lobata Platanus racemosa LOW TREE Fraxinus latifolia Juglans hindsii LIANA Rhus diversiloba Aristolochia calif. Smilax californica Clematis ligusticifolia SHRUBS Rubus vitifolius Rosa californica	Avena barbata Avena fatua Bromus diandrus Bromus mollis Carex (4 sp) Elymus glaucus Elymus triticoides Hordeum brachyantherum Hordeum geniculatum Hordeum leporinum Lolium multiflorum Lolium perenne Stipa pulchra

Elevation relative to river; time between major disturbances

Lo ←————————————————————————→ Hi

Frequency, duration, and depth of inundation

Hi ←————————————————————————→ Lo

52

Figure 3.   IDEALIZED TOPO-SEQUENCE OF HYDRIC VEGETATION PATTERNS -- SACRAMENTO-SAN JOAQUIN DELTA, CA.

OPEN AQUATIC	DENSE TULE ZONE	FRESH WATER MARSH	SHRUBS ON HUMMOCKS (in matrix of fresh water herbaceous marsh)	RIPARIAN THICKET OR FOREST (emerging from shrub layer on topographically higher positions)
Azolla filiculoides	Scirpus acutus	Polygonum	Cornus stolonifera	Populus fremontii
Elodea	S. californica	Typha	Rubus vitifolius	Salix lasiolepis
Myriophyllum	+ trace amount:	Scirpus acutus	Aonus rhombifolia	
Polygonum coccineum	Calistegia	Carex	Hibiscus californicus	
P. hydropiperoides	Galium trifidum	Lippia nodiflora		
Potomogeton pectinatus	Hydrocotyle umbellatum	Juncus		
Ludwegia peploides		Sagittaria latifolia		
	Low diversity - most	Verbena hastata		
	cover in Scirpus	Calistegia		
		Eleocharis macrostachya		
		Rumex crispus		
		Epilobium brevistylum var. ursinum		
		Lycopus		

53

Table 3. Density and basal area of trees greater than 1 dm. DBH. Glenn East
          site. n = 70

RIPARIAN FOREST -- GLENN EAST

	Density (trees/ha)	Relative Density	Basal Area (m^2/ha)	Relative B.A.	BA/tree (dm^2)
Populus fremontii		.44		.70	24.4
Quercus lobata		.07		.14	29.6
Salix sp.		.20		.08	5.8
Fraxinus latifolia		.19		.04	3.0
Platanus racemosa		.07		.04	9.2
Acer negundo		.015		<.001	.8
Cephalanthus occidentalis		.015		<.001	.8
TOTAL	252.1	1.0	38.6	1.0	15.3

Gravel Bar Thickets

Well-stabilized gravel bar deposits are dominated by sand bar willow (Salix hindsiana) (Figure 2) which forms dense thickets 3-5 m tall of up to 95% cover. Common associates include saplings of Alnus rhombifolia, Acer negundo, Fraxinus latifolia, and Populus fremontii, and shrubs of mule fat (Baccharis viminea). Scattered herbaceous species (Figure 2) are also present but their cover is generally low due to the dense canopy.

Open Floodplain Communities

Sand and gravel bars which are flooded annually support a sparse vegetation cover (5-25%) dominated by small (1m) shrubby and herbaceous perennials and annuals. The frequent disturbance normal to this habitat has favored invasion by many introduced species such as Bromus diandrus, B. tectorum, Salsola kali, Raphanus and Brassica spp, Tunica prolifera, Polypogon monspeliensis, and Verbascum thapsus. Native species of floodplains include the small shrubs Chrysopsis oregona, Trichostema lanatum, and Bidens laevis.

Hydric Communities

In old oxbows and the low areas a series of hydric communities occur. (Figure 3). Open water supports emergent and free-floating mat vegetation containing plants such as Polygonum spp., Ludwigia peploides, Azolla filliculoides, Potomogeton crispus, Elodea spp., and Myriophyllum spicatum ssp. exalbescens. Shallow water and low mud flats are dominated by Scirpus acutus (50-100% cover) 2-3 m tall. On higher areas, where Scirpus acutus is less prominent, the species diversity of the fresh water areas of the marsh increases considerably. Species such as Phragmites communis, Typha latifolia, T. domingensis, T. angustifolia, and Sparganium eurycarpum may occur here. Hummocks in higher areas of the marsh support shrub thickets of Cornus tolonifera, Cephalanthus occidentalis, and Rubus vitifolius with occasional Alnus rhombifolia and Fraxinus latifolia. It is also in this zone that the rare Hibiscus californica may be found. The Cornus and Cephalanthus hummocks are in turn invaded by species typical of the riparian forest: Alnus rhombifolia, Salix spp., Fraxinus latifolia, Rubus vitifolius, Rosa californica, as well as Populus fremontii. This turns higher hummocks into Alnus-dominated thickets and eventually Populus forests.

Summary and Conclusions

The riparian zone is a dynamic habitat. The vegetation of a given site reflects the history of flooding, aggradation, and erosion by the river. These habitats are subject to varying frequencies of flooding and of lateral erosion by the meandering river. The major riparian plant communities can be aligned along several topographic gradients (Figures 2 and 3). The low, recent gravel bar deposits are flooded frequently. Plant cover is low and is dominated by introduced annuals and low perennials. As gravel bars become more removed from the river and begin to stabilize, they are colonized by thickets of tall shrub and tree saplings generally dominated by Salix hindsiana. Riparian forest will become established on lower terrace deposits or as flood frequency decreases. These junglelike gallery forests are dominated by Populus fremontii and characterized by heavy cover of lianas. Higher ground in these forests supports Quercus lobata and Platanus racemosa. The older, higher terrace deposits support stands of valley oak woodland dominated by Q. lobata. These woodlands gradually thin out and grade into valley grassland vegetation with increasing distance from the river.

Oxbows and over-flow basins are characterized by a series of hydric communities. Fresh water marsh in low, wet areas is dominated by Scirpus acutus. On higher ground, this is succeeded by shrubs such as Cornus stolonifera and Cephalanthus occidentalis. These shrub-dominated habitats appear transitional to typical Populus fremontii dominated riparian forests on higher ground.

Literature Cited

Brumley, Terry D. 1976. Upper Butte Basin Study 1974-1975, State of California Resources Agency, Wildlife Management Branch. Admin. Report No. 76-1. 30 pp + Appendix.

Ceska, A. and H. Roemer. 1971. A computer program for identifying species-releve groups in vegetation studies. Vegetatio 23:255-276.

Cottam, Grant and J. T. Curtis. 1956. The use of distance measures in phytosociological sampling. Ecology 37(3):451-460.

McGill, Robert R., Jr. 1975. Land use changes in the Sacramento River riparian zone, Redding to Colusa. State of Cal. Resources Agency, Department of Water Resources. April, 1975. 23 pp.

Michny, Frank J., David Boos, and Frank Wernette. 1975. Riparian habitats and avian densities along the Sacramento River. Cal. Resources Agency, Dept. of Fish and Game. Admin Rpt. No. 75-1. March, 1975. 42 pp.

Mueller-Dombois, D. and H. Ellenberg. 1974. Aims and Methods of Vegetation Ecology. Wiley & Sons, N. Y. 547 pp.

Robichaux, Robert. 1977. Evolution of the Riparian Forest of California. In proceedings of "Riparian Forests in California: Their Ecology and Conservation." Symposium May 14, 1977. U. C. Davis.

Stebbins, G. Ledyard. 1971. Evolution and diversity of arid-land shrubs. In Wildland Shrubs - Their Biology and Utilization, USDA For. Serv. Tech. Rep INT-1, 1972. pp. 111-120.

Stone, Thomas B. 1976. Birds in riparian habitat of the upper Sacramento River. State of Cal. Resources Agency, Dept. of Fish and Game. Memorandum Report. Nov. 1976. 22 pp. + Appendix.

Stroud, Dennis. 1977. Comparative ecology of populations of Rattus rattus and R. norvegieus in a riparian habitat. Abstract. Field Studies in California. ERS Symposium. CSUS, Sacto. p. 1.

Thompson, Kenneth. 1961. The riparian forests of the upper Sacramento Valley. Ann. Assoc. Amer. Geog. 51(3):294-315.

West, James. 1977. personal communication.

Chapter 7

THE VALLEY RIPARIAN FORESTS OF CALIFORNIA: THEIR IMPORTANCE TO BIRD POPULATIONS

David A. Gaines
P. O. Box 886, Davis, CA. 95616

Introduction

Those who have heard the spring chorus of songbirds, watched herons feed their young in tree-top nests, glimpsed swarms of warblers in the early autumn greenery and tried to count wintering flocks of sparrows know first-hand the wealth and diversity of California's valley riparian forest avifauna. Today, with the last extensive remnants of these forests in jeopardy, it behooves us to weigh the importance of riparian habitat to birds and other wildlife (Gaines 1976).

Birdwatchers have long been aware of the "unparalleled diversity of the bird life" to be found in the broad-leaved deciduous hardwood forests which line the banks of lowland rivers (small 1974). Broad, qualitative impressions, however, are of limited value in documenting the importance of a habitat to bird populations. Ideally we should base our evaluation on a thorough knowledge of avian densities, foraging strategies, trophic dynamics, nesting requirements, and sensitivity to habitat alteration.

Despite their diverse avifauna, however, California's riparian forests have received scant attention from field ornithologists. The only discussions in the literature are those of Miller (1951), Small (1974), and Gaines (1974b).

This paper is based primarily on censuses conducted during the past five years at six riparian forest sites in the Sacramento Valley, California (Atwood 1976, Dembosz et al 1972, Gaines 1973, Laymon pers. comm., Manolis 1973, Metropulos 1974, Shuford 1973, Tangren 1971 and 1972, Winkler 1973a and 1973b). Most of this data was obtained using spot mapping or similar techniques (Enemar 1959, Svensson 1970). I have used it to quantify the density, diversity, foraging guilds, and other characteristics of the riparian avifauna.

The reader is cautioned against placing undue weight on the census data. Certain groups of birds, such as hawks (Falconiformes) and owls (Strigiformes), are usually not reliably censused. The small size of the plots (6-12 hectares) precludes accurate appraisal of colonial nesting species, such as Great Blue Heron and Purple Martin. The census sites may not be representative of riparian forest habitat elsewhere in the Sacramento and other California Valleys. Species whose populations fluctuate from year to year may have been misleadingly sampled. One recent study casts doubt on the accuracy of the mapping method itself (Best 1975). Nevertheless these censuses provide the best description of the riparian forest avifauna possible at the present time.

Although the data derives entirely from the Sacramento Valley, the discussion is generally applicable to the birdlife inhabiting valley riparian forests throughout cismontane California. With the exception of those along the Colorado River, these forests resemble one another in terms of plant species composition, physiognomy of the vegetation, and avian inhabitants. The Colorado River forests, with their affinities to the Sonoran flora and fauna of northern Mexico, are excluded from this survey.

Data from valley riparian forest outside the Sacramento Valley is limited to a single breeding bird census near Fresno, Fresno County (Ingles 1950) and a winter bird population study near Riverside, Riverside county (Hay 1976). Future field work will broaden the data base. Additional distributional information was derived from the species accounts in Grinnell and Miller (1944), the Middle Pacific Coast Region Reports in American Birds (formerly Audubon Field Notes), and the unpublished notes of individuals acknowledged in the text.

This survey is divided into five sections: (1) a brief description of the habitat, (2) a discussion of the breeding avifauna, (3) a discussion of the wintering avifauna, (4) a discussion of migration, and (5) a reflection on the plight of the Yellow-billed Cuckoo.

This paper is descriptive, not theoretical. Many pages could be written on niche relationships, competitive regimes, optimization strategies, and other aspects of community structure. However interesting these topics might be, the importance of riparian forest habitat to bird populations is of greater concern. Unless we can demonstrate the value of preserving these forests, theoretical discussion will be purely academic.

Forests of Tropical Luxuriance

The term "riparian forest or woodland" is applied by most botanists to the broad-leaved and winter deciduous, phreatophytic tree formation that lines watercourses (Ornduff 1974). On the elevated natural levees or rimlands that line the banks of large, aggrading streams, such as the Sacramento River, these hardwood forests rival in grandeur the coniferous forests for which California is famous. Nowhere else in the lowlands of the arid west does one encounter "fine jungles of tropical luxuriance" (Muir 1894).

The vegetational characteristics of riparian forests vary with the frequency and duration of inundation by floodwater. Gravel bars and sandbars remain submerged for so long a period that only a few shrubby, flood-battered willows will grow there. These are pioneers, so to speak, from the dense cottonwood-willow (Populus-Salix) forests which thrive on slightly higher ground. Where flooding is still less frequent, valley oak (Quercus lobata) forests replace cottonwood-willow as the dominant vegetation. If these oaks are removed, however, cottonwoods and willows may re-colonize. In this sense, the latter is a seral and the former a climax vegetation type. Cottonwood-willow and valley oak riparian forests may be thought of as arbitrary stopping points along gradients of decreasing moisture and disturbance. They differ so significantly in terms of birdlife, however, that it is useful to consider each independently.

From the avian perspective the most important aspect of the cottonwood-willow type is its stratified foliage profile. A closed canopy of tall, mature cottonwoods and tree willows provides a habitat niche for a shade-tolerant understory of younger or smaller trees, shrubs, vines, and forbs. Among these plants, blackberry (Rubus spp) and wild grape (Vitis californica) produce fruits that are important seasonal food sources for some of the birds.

In the valley oak type the canopy is usually open. Large, gnarled oaks are dispersed in groves with intervening grass and forb-covered openings. A shrubby understory provides cover for birds which forage on or near the ground. The seeds of herbaceous plants, oak acorns, and the fruits of poison oak (Rhus diversiloba), blue elderberry (Sambucus mexicana), and the parasitic mistletoe (Phoradendron flavescens) are important seasonal food sources.

Four of the six riparian forest census sites fall under the cottonwood-willow type, two under the valley oak type. Because the latter are mixed with cottonwoods, I have labelled them "oak-cottonwood" on the tables and figures. For more detailed habitat descriptions the reader is referred to the quantitative vegetation analyses included in the nine published censuses (see lists of sources accompanying Tables 1 and 2).

The Breeding Avifauna

Those birds known to nest or to have nested historically in the riparian forests of the Sacramento Valley are listed in Table 1. Most breed in similar habitat throughout cismontane California (Belding 1890, Grinnell and Miller 1944, Gaines 1974b).

The percentage of breeding individuals which are migratory differs strikingly between the cottonwood-willow and oak-cottonwood census plots (Figure 1). In the former a large influx of birds which winter in subtropical areas, such as Western Wood Pewee, Yellow Warbler, and Northern (Bullock's) Oriole, account for 36% of the nesting bird density. In the valley oak forest, in contrast, only 4% of the nesting birds are migratory. Moister conditions in the cottonwood-willow forests may promote lusher plant growth, higher invertebrate populations and, therefore, more available food for flycatchers, warblers, and other migratory, insectivorous birds.

Based on Miller's (1951) analysis of the California avifauna, 43% of the species and 38% of the individuals breeding in cottonwood-willow habitat have a "primary affinity" to riparian forest (Table 1). In other words, in comparison to 21 other California vegetation types, these forests probably support the highest concentrations of these species. In cismontane California Red-shouldered Hawk, Yellow-billed Cuckoo, Willow Flycatcher, Bell's Vireo, Yellow Warbler, Yellow-breasted Chat, and Blue Grosbeak breed in no other forest habitat.

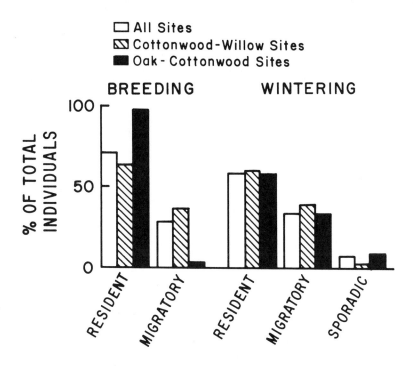

Figure 1. Status of breeding and wintering birds censused at six riparian forest sites in the Sacramento Valley, California. The histograms depict the percentage of total individuals which are resident, migratory or sporadic for (1) all sites, (2) cottonwood-willow sites and (3) oak-cottonwood sites.

The breeding avifauna of California's riparian forests has intriguing affinities to that of the similarly winter-deciduous hardwood forests of eastern North America (Miller 1951). Many typically "Eastern or Mid-eastern" species, such as Red-shouldered Hawk, Yellow-billed Cuckoo, Downy Woodpecker, Bell's Vireo, Warbling Vireo, Yellow Warbler, Yellow-breasted Chat, Blue Grosbeak, American Goldfinch, and Song sparrow, have been able to colonize the arid West primarily because humid, broad-leaved riparian forests offered congenial haunts.

Interestingly, all of these birds have evolved western subspecies (American Ornithologist's Union 1957). Three of these races, the Red-shouldered Hawk (Buteo lineatus elegans), the Bell's Vireo (Vireo belli pusillus), and the Blue Grosbeak (Guiraca caerulea salicaria), breed only in the valleys of California.

The average density of nesting birds on the cottonwood-willow census plots (2088/km^2) is strikingly higher than that on the oak-cottonwood plots (1279/km^2) (Table 1). This difference is due primarily to migratory species. If we only consider residents, the density in cottonwood-willow (1336/km^2) is only slightly higher than that in oak (1227/km^2).

Breeding bird densities in cottonwood-willow forests equal or exceed those in any California vegetation type for which census data is available (Gaines 1974b). The dense, stratified cottonwood-willow forest vegetation may facilitate high breeding bird densities. With increased trunk, branch, and foliage foraging space, bird territories may occupy less ground surface area.

The large number of migratory birds implies a seasonal abundance of insect food during the warmer months (DeSante 1972). A recent study, however, suggests that bird densities do not depend on habitat productivity (Willson 1974). In this regard it would be interesting to try to correlate bird densities in riparian forest habitats with plant productivity and invertebrate populations.

California's riparian forests support a high diversity of breeding birds (Miller 1951).

Excluding Ring-necked Pheasant and Western Meadowlark (included because some census plots edge on grassland) 67 species are known to nest in the forests of the Sacramento Valley (Table 1). Species richness (number of species) equals or exceeds that in any habitat for which census data is available (Gaines 1974b).

Most avian ecologists think of diversity not only in terms of species richness, but also in terms of the relative abundance of one species to another. Thus a community of 10 species of equal density is considered more diverse than a community of one common species and nine rare ones. The Shannon formula provides an index of both of these components (Shannon and Weaver 1963):

$$\text{Species Diversity} = -\sum_{i=1}^{s} p_i \ln p_i$$

where p_i equals the proportion of the ith species to the total population and s equals the total number of species.

Using this index the average species diversity for the cottonwood-willow census plots (3.17) is considerably higher than that for the oak-cottonwood plots (2.51) (Table 1). Species richness, however, is only slightly higher (27 to 24). Thus the high diversity values in cottonwood-willow reflect a large number of species with relatively even densities.

This high diversity seems to depend, not on edge effect or plant species diversity, but on foliage volume and foliage height profile. One of the most interesting census results is the lack of correlation of diversity with the extent that riparian forest habitat edges on openings or other types of vegetation. Most species are more or less evenly dispersed within the forest, with little or no tendency to concentrate near the edge (DeSante 1972). Thus the theory that diversity is enhanced by the mixture of species from adjacent habitats may not apply to riparian forests.

Beginning with MacArthur and MacArthur (1961) a series of studies have linked bird species diversity in forest communities with foliage height diversity, foliage volume, and other habitat characteristics. This complex, fascinating subject has recently been summarized by Balda (1975). In addition to foliage, such factors as food resources, nest sites, nesting material, song posts, proximity to water, extent of habitat, geologic history, and human disturbance need to be considered. Understanding these factors is important to assuring a diverse avifauna in sanctuaries, state parks, and other lands set aside as riparian forest preserves.

Foraging Guilds

In order to better describe how birds utilize the riparian forest environment, I have grouped species into foraging and nesting guilds (Table 1). Root (1967) defines a guild as "a group of species that exploit the same class of environmental resources in a similar way." I have followed Salt (1953) in classifying species into foraging guilds according to the type of substrate on which they forage, and the major component of their diet. Downy Woodpecker, Nuttall's Woodpecker, Plain Titmouse, and White-breasted Nuthatch, for example, feed primarily on invertebrates gleaned or drilled from the bark of trees. They are thus grouped together in the "bark insect" foraging guild.

Since no studies have been published on the behavior and diets of California's riparian forest birds, I have based my foraging guild taxonomy on data presented by Bent (1963) and personal observations. In some cases a single species has been placed in two guilds. The Rufous-sided Towhee, for example, is classified in the "ground seed" and "ground insect" categories. For purposes of analysis in these cases, each of the two guilds in weighted equally. A few species, such as the Scrub Jay, are considered "generalist omnivores." These opportunistic birds forage from the ground to the canopy, and feed on whatever is available. The "seed" guilds include both granivorous and frugivorous birds. Classifications are based on behavior and diet during the nesting season and thus may not apply to other times of the year.

The percentage of total breeding individuals in each of ten foraging guilds is graphed in Figure 2. Because the analysis is based on densities and not on biomass, the importance of large birds such as hawks and owls (ground mammal and foliage bird predators) is underestimated. A study in Sacramento County came to the conclusions that "riparian habitat was used by more raptors than any other habitat" and that "nesting success was greatest" there (Vincent 1974).

Figure 2. Foraging guilds of breeding birds censused at four riparian forest sites in the Sacramento Valley, California. The histograms depict the percentage of total breeding individuals in each guild for (1) all sites, (2) cottonwood-willow sites and (3) oak-cottonwood sites.

The oak-cottonwood census plots are dominated by generalist omnivores (32%), such as Scrub Jay and Starling. The cottonwood-willow plots are dominated by foliage insectivores (36%), such as Bewick's Wren, Northern Oriole, and Black-headed Grosbeak. If the introduced Starling is excluded, however, the foliage insect and ground seed guilds assume primary and secondary importance in both forest types.

As might be expected in a forest environment, over half the birds forage in the foliage or on the bark of trees. More surprising is the large complement (27%) of ground seed and/or insect eaters such as California Quail and Rufous-sided Towhee. A dense understory of shrubs, grasses and forbs, and an insect-rich duff provide cover and food for these inhabitants of the forest floor.

As already noted, cottonwood-willow riparian forests support a higher density of breeding birds than do valley oak forests, primarily because resident nesting populations are augmented by migrants. Most of these migratory birds belong to the foliage insect (47%) or air insect (34%) foraging guilds. This is not surprising; species wedded to foliage, such as Yellow-billed Cuckoo, Yellow Warbler, and Northern Oriole, must depart before the autumn leaf fall. Likewise, species specialized to capturing insects on the wing, such as Western Wood Pewee and Tree swallow, must depart before the frosty days of fall threaten their food supply.

Interestingly, the foliage insect guild is about equally divided between resident birds (53%) and migrants (47%). These residents, such as Bewick's Wren and Bushtit, are sufficiently resourceful in foraging habits to switch guilds during the colder months.

Resident birds dominate the ground seed or insect (86%), bark insect (100%), and generalist omnivore (100%) foraging guilds. In contrast to winter-deciduous foliage, ground and bark substrates are available throughout the year. Ground feeding birds, such as California Quail and Rufous-sided Towhee, and bark insectivores, such as Nuttall's Woodpecker, Downy Woodpecker, and White-breasted Nuthatch, are thus able to winter in their breeding haunts. The opportunistic generalist omnivores are able to find food in riparian forest habitat at all seasons.

The following generalizations derive from the above discussion and Figure 2:

1) Foliage, ground, and bark all provide important foraging substrates for nesting riparian forest birds.

2) Insects are the primary food source of most species, although seeds also are important, especially for ground foraging birds.

3) Seasonal foliage and insect abundance permit a large number of migratory birds to breed in cottonwood-willow riparian forest habitat.

Nesting Guilds

In order to reproduce successfully, birds require not only an adequate supply of food for themselves and their young, but also a suitable place to nest. Riparian forest species may be grouped into the following four guilds based on their usual choice of nesting site:

1) Birds which nest on the ground (such as California Quail and Rufous-sided Towhee).

2) Birds which nest in shrubs (such as Yellow-breasted Chat and Song Sparrow).

3) Birds which nest among the foliage or branchwork of trees (such as Great Blue Heron, Western Wood Pewee, Scrub Jay, and Black-headed Grosbeak).

4) Birds which nest in tree holes or cavities (such as Wood Duck, Screech Owl, Downy Woodpecker, Ash-throated Flycatcher, Plain Titmouse and Starling).

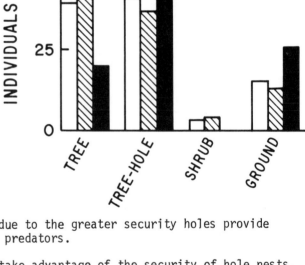

Figure 3. Nesting Guilds of breeding birds censused at four riparian forest sites in the Sacramento Valley, California. The histograms depict the percentage of total breeding individuals in each guild for (1) all sites, (2) cottonwood-willow sites and (3) oak-cottonwood sites.

The percentage of total breeding individuals in each of these four nesting guilds is graphed in Figure 3. As would be expected in a forest environment, most birds (84%) nest above the ground in woody vegetation. Worthy of special attention is the high percentage (41%) utilizing tree holes or cavities.

The data summarized by Welty (1975) clearly shows the superior success of hole-nesting verse open-nesting species both in hatching eggs and in fledging young. This is due to the greater security holes provide from squirrels, raccoons, jays, and other nest predators.

One may well ask why more species do not take advantage of the security of hole nests. In fact these sites may be in such short supply that they limit the density of cavity-nesting birds. Pfeifer (1963), for example, was able to increase a forest bird population density 25-fold by erecting nest boxes and other artificial nesting facilities. Studies of the hole-nesting Eastern Bluebird (Sialia sialis) and Pied Flycatcher (Ficedula hypoleuca) have correlated population density with the availability of nesting sites (Laskey 1940, von Haartman 1956).

Most of the tree-hole nesting sites in riparian forests are excavated by woodpeckers in the soft wood of dead, decaying trees and snags. Since woodpeckers carve out a new domicile each year, their former abodes are available to other cavity-nesting species. Thus the preservation of dead trees and snags is an important consideration in managing riparian forest habitat for bird populations. Only 16% of riparian forest hole-nesting birds are migratory. Resident species probably have the advantage in laying early claim to these desirable nesting sites.

Riparian forests provide breeding and roosting sites for nine species of water birds which forage in surrounding marsh and riverine habitats. Two fish-eating raptors, Osprey and Bald Eagle, build their bulky stick nests in trees near the rivers where they hunt. Two species of waterfowl, Wood Duck and Common Merganser, raise their young in tree cavities. Of particular importance are colonial nesting rookeries of Great Blue Heron, Great (common) Egret, Snowy Egret, Black-crowned Night Heron, and Double-crested Cormorant.

The largest heron rookery for which reliable data is available was discovered by Emlen (1938) along the Sacramento River in June, 1937. A thousand nests, most of them occupied, were situated in the tops of tall cottonwood trees about three miles below Ord, Butte County, on the Llano Seco Rancho. Emlen tallied 400 pairs of Great Egrets, 200 pairs of Great Blue Herons, 150 pairs of Double-crested Cormorants, and 80 pairs of Black-crowned Night Herons. This heronry was jeopardized by logging operations in 1975 (Gaines 1976).

Hundreds of pairs of herons and egrets flying to and from their bulky, tree-top nest are an impressive sight. Usually the nests are ensconsed high in the branches of a sturdy valley oak or sycamore, but sometimes cottonwoods are utilized. The nests are re-occupied year after year.

Avifaunal Changes and Human Impact

Since white settlement of the valleys of California little more than a century ago, the nesting riparian forest avifauna has changed significantly. The following accounts survey species for which meaningful data is available:

Double-crested Cormorant

Grinnell and Miller (1944) reported "marked reduction in numbers of individuals and breeding colonies." At least one cormorant colony is known to have succumbed to a gunner who "liked fishing and considered them competitors" (Moffitt 1939). The lack of nesting pairs in the Sacramento Valley is symptomatic of an alarming decline throughout the inland valleys of California (Gaines 1974b). According to Belding (1878), cormorants were formerly "abundant at all seasons, particularly where sloughs penetrated the oaks of the uplands."

Great Blue Heron and Great Egret

The available evidence suggests that heron rookeries were formerly much more numerous in riparian forests, at least in the Sacramento Valley. Storer (1933), for example, was told by an early resident of Yolo County that "during his boyhood days in the late '70's or early '80's, herons nested 'by the hundreds' in groves of trees then standing along the banks of Putah Creek, between the present site of the University Farm and the present Dixon-Woodland highway." None at all presently breed along Putah Creek, nor anywhere else in southern Yolo County. The following report, one of many in the literature, describes one cause of the decline: "a Great Blue Heron and Great Egret rookery west of Gridley which held 600 nests five years ago has been completely wiped out because of the trees being cut and broken down" (Cogswell 1956). Encroachment on the colonies of these shy birds, together with the loss of wetland habitat, continues to reduce their breeding populations in inland valleys throughout cismontane California.

Cooper's Hawk

According to Dawson (1923), the Cooper's was the "most abundant" hawk in California; Grinnell and Miller (1944) judged it "varyingly common." The paucity of nesting records in recent years is cause for concern (Arbib 1976).

Bald Eagle

This species was formerly "common" along the rivers of lowland California (Belding 1878, Grinnell 1915). No Bald Eagles are presently known to nest in valley riparian forests.

63

Yellow-billed Cuckoo

The Yellow-billed Cuckoo, formerly "fairly common" (Grinnell 1915) in cottonwood-willow forests, has become so scarce in California that it has been added to the official state list of "rare and endangered" wildlife (Gaines 1974a). This species is discussed in detail in the concluding section of this paper.

Willow (Traill's) Flycatcher

Belding (1890) considered this species "a very common summer resident in willows of central California, most so along the valley rivers." Breeding Willow Flycatchers have vanished from almost all of their former riparian forest haunts in the valeys of California (Gaines 1974b, Table 1).

European Starling

The non-native Starling has spread west from New York, where it was introduced in the last century (Bent 1963). Since it was first recorded in California in 1942 (Jewett 1942), it has increased phenomenally in numbers. Although Starlings forage primarily in agricultural fields and orchards, they resort to tree holes for nesting sites. The result is a large influx into riparian forest habitat during the breeding season. In fact, densities of Starlings were higher than those of any other species on four of the eight breeding bird censuses (Table 1).

Bell's Vireo

"This very interesting little bird," wrote Belding in 1879, "is common in summer in willow thickets. . . it is active, restless, noisy or musical, and does not fail to make its presence known." Grinnell and Miller (1944) considered the Bell's Vireo "common, even abundant locally." During the past 20 years, however, no breeding pairs have been found anywhere in the riparian forests of cismontane California (Cogswell 1958, Gaines 1974b). Hence the race Vireo bellii pusillus, which nests only in the valleys of California, is probably on the brink of extinction.

Warbling Vireo

This vireo was considered "common" in the valley riparian forests of California by Belding (1890) and Grinnell (1915). In recent years, however, none have nested in the Sacramento Valley. They have likewise been absent from other areas of seemingly suitable valley riparian forest habitat (Gaines 1974b, Table 1).

Yellow Warbler

This species, formerly a "common" breeding bird in the valley riparian forests of California (Grinnell and Miller 1944), has declined throughout its lowland range (Arbib 1976). In 1924, for example, Grinnell (1924) considered the Yellow Warbler and the Bell's Vireo the "usual companions in streamside willows" along the Sacramento River. In recent years, however, neither of these species have bred in many areas of seemingly suitable habitat (Gaines 1974b, Table 1).

Common Yellowthroat

Grinnell and Miller (1944) considered this species a "common" nesting bird in riparian willow thickets. In recent years, however, Yellowthroats have failed to breed in many areas of seemingly suitable habitat (Gaines 1974b, Table 1).

Brown-headed Cowbird

Until 1900 a single bird in the Sacramento Valley remained the only record of the Brown-headed Cowbird from cismontane California (Grinnell 1915). But with the spread of feedlots and irrigated agriculture, they arrived in flocks. In 1933, for example, Willet wrote that "the increase in numbers of the cowbird in Southern California during the past 20 years has been remarkable; in fact, unparalleled by any other of our native birds." The same story repeated itself throughout the state (Grinnell and Miller 1944). During recent years, flocks of up to 10,000 individuals have been counted in the Sacramento Valley (Betty Kimball, pers. comm.). In the breeding season, cowbirds invade riparian forest habitat, where they burden other species with the task of incubating their eggs and raising their young (Figure 4).

Figure 4. Bell's Vireo feeding a nestling Brown-headed Cowbird. The decline of
Willow Flycatcher, Bell's Vireo, Warbling Vireo, Yellow Warbler and
Common Yellowthroat has been attributed to cowbird parasitism (see text).
Drawing by Cameron Barrows.

Thus, during the past fifty years, the breeding populations of at least 11 species
have declined strikingly in California's valley riparian forests. Black-crowned Night
Heron, Long-eared Owl, Swainson's Thrush, Blue-gray Gnatcatcher, Yellow-breasted Chat
and Song Sparrow should probably be added to this list, although the evidence is not
entirely conclusive.

Two species not originally present, Starling and Brown-headed Cowbird, now constitute
a significant percentage (9%) of the breeding bird density.

Dwindling numbers of Cooper's Hawks and Yellow-billed Cuckoos are at least partially
attributable to habitat attrition, for both require extensive territories before they
will nest. The plight of Willow Flycatcher, Bell's Vireo, Warbling Vireo, Yellow Warbler
and Common Yellowthroat, however, cannot be blamed on loss or alteration of habitat.
Throughout the valleys of cismontane California, these species have vanished from many
seemingly unaltered haunts where they were formerly numerous.

The troubles of Bell's Vireo and other small, nesting riparian forest passerines has
been attributed to another bird, a newcomer, the Brown-headed Cowbird (Pray 1953, Gaines
1974b). Once a female cowbird lays her eggs in the nests of vireos, warblers, or other
susceptible hosts, she takes no further interest in her progeny. The hosts hatch her eggs
and raise her young at the expense of some of their own brood.

The case against the cowbird rests on (1) their recent spread to California and (2)
their preferences for certain hosts. Parasite-host relationships, such as that between
the cowbird and the passerines it parasitizes, are usually detrimental only when the

65

interacting populations have not had a common evolutionary history. Given such a history, natural selection tends to moderate the impact of the parasite to a level bearable by its host, since the former is dependent on the latter for survival.

The cowbird, however, is a generalist parasite that does not specialize on any one host. If other nesting species are available, pressure on a given host, such as the Bell's Vireo, may not relent even when its numbers drop drastically. Hence where cowbirds are numerous, hosts must evolve behavioral defenses or adjustments in reproductive rates to maintain their populations. In California, however, the spread of agriculture has "allowed the cowbird to penetrate into new regions where it has access to host populations that have had little or no ancestral experience through which to develop effective defenses against it" (Mayfield 1965).

In the 1920's Hanna (1928) studied the influx of cowbirds into riparian forest habitat in the San Bernardino Valley of Southern California. Willow Flycatcher, Bell's Vireo and Yellow Warbler were, in that order, the cowbird's most frequent hosts. In the nearby San Gabriel Valley, Rowley (1930) came to similar conclusions. According to Dawson (1923), "it is rare to find Bell's Vireos' nests which have not been victimized, and the destruction caused to this one species is enormous." Based on information amassed by Friedmann (1929 and 1963), Willow Flycatcher, Bell's Vireo, Warbling Vireo, Yellow Warbler and Common Yellowthroat are precisely those riparian passerines most frequently parasitized.

Of the birds which have declined, only the Bell's Vireo has seemingly vanished entirely. Perhaps the other species have managed to maintain themselves in valley riparian forests through accretion from neighboring, better insulated populations-- for instance, in the case of the Yellow Warbler, those of mountain canyons. The range of the Bell's Vireo, in contrast, is restricted to lowland riparian where pressure from cowbirds is greatest.

Still, the evidence is inconclusive. During the 1950's, for example, Yellow Warblers, Lucy Warblers and Bell's Vireos declined along the Lower Colorado River (Monson 1960), an area where Brown-headed Cowbirds have been "common" throughout the century (Grinnell 1915). Before we condemn the cowbird, more information is needed on how they affect the population dynamics of their hosts. The few studies suggest that of the 14 or more eggs each female lays in a season, 25-35% produce fledglings. These young cowbirds are raised at an average expense of one of the host's brood (Nice 1937, Norris 1947). Specific species or local populations, however, may bear higher pressure. Of the Red-eyed Vireos (Vireo olivaceus) studied by Southern (1958) 73% were parasitized, and the total population raised, on the average, only one of their own young per pair. These studies pertain to eastern North America. Information specific to California is needed. In the case of the Bell's Vireo, this may be critical to saving the California subspecies.

The other avian newcomer, the European Starling, may also be affecting breeding bird populations by competing for tree hole nesting sites. If these sites are limiting to hole-nesting birds, then one would expect the increase in breeding Starlings to have been accompanied by a corresponding decrease in hold-nesting native birds. The few published studies, however, are inconclusive. Troetschler (1976) concluded that "increased need for hole defense. . . may limit Acorn Woodpecker breeding in the future" and that "breeding Red-shafted Flickers may be found primarily in high mountains and forests." The disappearance of nesting Common (Red-shafted) Flickers and Tree Swallows from riparian forests along Putah Creek between Davis and Winters, Yolo County, is probably attributable to Starlings (Gaines, unpublished data). Ash-throated Flycatchers, Tree Swallows, Plain Titmice, White-breasted Nuthatches and House Wrens may be particularly susceptible because they do not excavate their own nest holes. Understanding the impact of Starlings may thus be critical to preserving a diverse hole-nesting avifauna in valley riparian forest habitat.

The Wintering Avifauna

Those birds known to winter in the riparian forests of the Sacramento Valley are listed in Table 2. Most winter in similar habitat throughout cismontane California (Grinnell and Miller 1944).

A factor which complicates interpretation of the census data is the sporadic occurrence of Band-tailed Pigeon, Varied Thrush, Golden-crowned Kinglet, Purple Finch, and other irruptive montane birds (Figure 1). During some years one or more of these species may comprise a significant proportion of the wintering avifauna. In other years they may be rare or absent. In February, 1973, for example, 5000 Band-tailed Pigeons were tallied in a 12 mile stretch of riparian forest along the Sacramento River in Yolo and Sacramento Counties (Richard Stallcup, pers. comm.); most years none winter in this area. These "irruptions" of sporadic species into lowland forests are thought to correlate with food shortages in their usual montane winter haunts (Svardson 1957).

Not all year-to-year variations in wintering bird populations are attributable to "sporadics." A comparison of the 1975 (1A) and 1976 (1B) Dog Island census data, for example, shows a ten-fold increase in Bushtits and a 12-fold decrease in American Robins. Because of this variability, it is difficult to draw conclusions from a data base of only four censuses.

Nonetheless it is interesting to note the high percentage of migratory or sporadic birds (40%) on both the cottonwood-willow and valley oak census plots (Figure 1). This suggests that riparian forests are important wintering areas for Sharp-shinned Hawks, Hermit Thrushes, Ruby-crowned Kinglets, Yellow-rumped Warblers, Dark-eyed (Oregon) Juncos, Golden-crowned Sparrows and other birds which breed in montane and northern coniferous forests. The availability of wintering habitat may be a factor limiting the populations of some of these species (DeSante 1976).

The average density of wintering birds on the valley oak plots ($2439/km^2$) is strikingly higher than that on the cottonwood-willow plots ($997/km^2$) (Table 2). It is interesting to compare these figures with breeding bird densities (Table 1). The data suggest that oak forests support 90% more wintering than nesting birds, and cottonwood-willow forests almost the reverse. This same trend, although less pronounced, is reflected by the data on species richness and diversity.

These seasonal changes are due primarily to migrants. The large number of breeding birds which leave cottonwood-willow forests before the autumn leaf-fall deplete wintering bird densities. In the oak forests, in contrast, a large influx of migratory wintering species augments the largely resident breeding population.

Most (69%) of these migrant birds subsist on seeds and/or fruits (Table 2). More open conditions in the oak forests may promote the growth of herbaceous, seed-producing forbs and grasses. Berry-producing plants are probably more abundant. The available census data suggest that average bird density in oak riparian forest exceeds that in coastal mixed forests, coastal coniferous forests and chaparral (Stewart 1972).

Wintering bird diversity is also high. Excluding Western Meadowlark (included because a census plot edges on grassland) 60 species are known to winter in the riparian forests of the Sacramento Valley (Table 2). The census data misleadingly suggest that average species richness is higher in winter (42 species) than in summer (26 species) (Tables 1 and 2). Methodology is probably responsible for most, if not all, of this difference. Wintering censuses count all birds, whereas breeding censuses count only territorial males. In fact, slightly more species are known to breed in riparian forests than are known to winter, at least in the Sacramento Valley (Tables 1 and 2).

Despite their greater species richness, average species diversity ($- \Sigma\ p_i\ \ln\ p_i$) on the wintering bird plots (3.01) is scarcely higher than that on the breeding bird plots (2.92) (Tables 1 and 2). This reflects the relative abundance of one species to another. On the wintering plots a few species, such as California Quail, Scrub Jay, Yellow-rumped Warbler and Golden-crowned Sparrow, are much more numerous than most of the others. On the breeding plots, in contrast, species densities are distributed more evenly.

Foraging Guilds

The percentage of total wintering individuals in each of ten foraging guilds is graphed in Figure 5.

Wintering birds must find food in cold, foggy or rainy weather when active insects are scarce or non-existent. During these periods they subsist primarily on (1) quiescent insect life, such as eggs and pupae, (2) seeds, such as those of grasses and forbs, and (3) fruit, such as valley oak acorns, poison oak berries and mistletoe berries.

Most wintering birds are plastic enough in their foraging habits to take advantage of

67

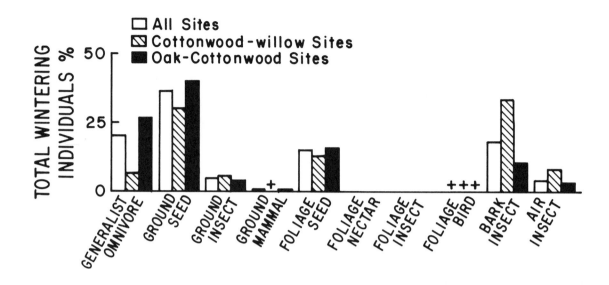

Figure 5.   Foraging guilds of wintering birds censused at three riparian forest
            sites in the Sacramento Valley, California. The histograms depict the
            percentage of total wintering individuals in each guild for (1) all sites,
            (2) cottonwood-willow sites and (3) oak-cottonwood sites.

different kinds of food that become available as weather conditions change. On a balmy day,
for example, Ruby-crowned Kinglets and Yellow-rumped Warblers, which usually glean insects
from bark, will sally for airborne insects in the manner of flycatchers. Even woodpeckers
sometimes engage in this atypical behavior.

With the change in seasons the foliage-gleaning insectivores of the breeding avifauna
are replaced by the ground-foraging granivores of the wintering avifauna. California Quail,
Rufous-sided Towhee, Dark-eyed Junco, White-crowned Sparrow, Golden-crowned Sparrow and
other ground-dwelling species comprise 43% of wintering bird density on the census plots.
These species depend on understory shrubs and brambles for escape and roosting cover.

Because of the lack of foliage, arboreal birds subsist primarily on bark insects and/or
fruits. Woodpeckers, such as Nuttall's and Downy, drill into trunks and branches in search
of larval or pupating insects. Bark insect gleaners, such as White-breasted Nuthatch and
Ruby-crowned Kinglet, pick their prey from trunks and limbs. Fructivorous American Robins
and Cedar Waxwings descend in flocks to feast on ripening berries.

Migration

Large numbers of passerine birds forage and shelter in riparian forest habitat during
their migratory journeys. Most are foliage-gleaning or sallying insectivorous species which
winter in subtropical Mexico and Central America. During the spring migration, these birds
pass northward on a broad front through the forests and woodlands of lowland California.
The hills are green, and deciduous foothill oaks have just leafed out, and insect life is
everywhere abundant. By late summer, however, the long dry period has seared the hills to
golden brown. At this season riparian forests provide the only lush, insect-rich forest
habitat in lowland, cismontane California. The importance of these forests to southward
(fall) migrants cannot be underestimated.

Until the early 1960's this southward migration was overlooked by field ornithologists.
During the sultry days of August and September they were too eager to reach the cool forests
of the higher mountains to stop and survey those along valley watercourses. The 1964
Audubon Field Notes report (Chase and DeBenedictis 1965) first called attention to the mag-
nitude of this lowland migration. "In the last four years," reads the report, "Western
Flycatcher, Orange-crowned Warbler, MacGillivray's, Nashville and Wilson's Warblers and
Western Tanager have been found in the period August 4-9 without fail. . . the presence of

large numbers seems to be indicated by the observations of 8/6/64. . . 173 migrants of seven species (including 75 Empids and about 50 Nashville Warblers) were seen in about one-fourth mile of riverside vegetation. . . by middle August numbers equivalent to this may be seen . . . and in September large numbers of migrants, including several species (as Yellow Warbler) not present in August, are readily found." ("Empids" refers to Empidonax Flycatchers, ed.).

Unfortunately, quantitative data on landbird migration in California has yet to be gathered. Most birdwatchers agree that the density and diversity of migrants in riparian forest exceed that in any other California habitat with the possible exception of the fir forests of the mountains.

In order to better indicate the magnitude of this migration I have included data on peak migratory bird densities in a cottonwood-willow forest in Tehama County (Stephen Laymon, unpublished data; Table 3). Less methodical censuses of a similar, but more open forest along Putah Creek in Yolo County between 1972 and 1976 differ significantly in species composition (Gaines, unpublished data). At the latter locality, for instance, Swainson's Thrush was a regular northward migrant during May, and Orange-crowned Warbler, Common Yellowthroat, Wilson's Warbler, and Western Tanager were most numerous as southward (fall) migrants. These scant data suggest that there is geographic variation in the size and species composition of migratory bird populations even in seemingly similar riparian forest habitats.

Banding records provide another source of information. During 1970 the Point Reyes Bird Observatory sponsored a project called "Operation Transect" which included banding of migratory landbirds in a riparian cottonwood-willow forest near Sacramento (Robert Stewart, unpublished data). Of 182 southward migrants, 65% were hatching year (immature) birds. Subcutaneous fat varied from 0 (none visible) to 2 (furcular cavity filled). This suggests that migratory birds lay-over several days in riparian forest habitat while they forage and store the fat needed to fuel their long, nocturnal, migratory journeys.

In order to more fully assess the importance of riparian forest habitat to migratory bird populations, more information is needed on age and sex distribution, turn-over rates, lay-over times, rates of weight and fat increase, densities, and habitat utilization.

The Yellow Billed Cuckoo

Nothing better illustrates the destruction of riparian forest habitat than the decline in California's populations of the Yellow-billed Cuckoo. This sinuous bird is closely restricted to broad expanses of cottonwood-willow forest. In the early part of this century the clearing of these forests was recognized as a threat to the cuckoo's survival (Jay 1911). At that time, they were still "fairly common" (Grinnell 1915). Only three decades later, however, Grinnell and Miller (1944) concluded that "because of removal widely of essential habitat conditions, this bird is now wanting in extensive areas where once found." Recent studies have confirmed this gloomy picture. Only in the relatively large remnants of forest that hug the Sacramento River between Colusa and Red Bluff are a few pairs still known to nest within cismontane California (Gaines 1974b).

The original distribution of the Yellow-billed Cuckoo in California reflects the pristine extent of its riparian forest habitat. According to Grinnell and Miller (1944), cuckoos nested in most of the coast range valleys from San Diego County north through the San Francisco Bay region as far as Sebastopol, Sonoma County and through the San Joaquin and Sacramento Valleys from the vicinity of Bakersfield, Kern County north to Redding, Shasta County.

Over most of this area, once extensive riparian forest habitat has been sacrificed to to civilization. The Santa Ana River in the San Bernardino Valley of Southern California is an excellent example. Here the Yellow-billed Cuckoo was first discovered nesting in California by Stephens in 1882 (Bendire 1895). During the 1920's Hanna (1937) found 24 nests in the "miles of cottonwood and willow" watered by the river. "In contrast with those good old days," he writes, "we now have very little water in Warm Creek and seldom any surface water in the Santa Ana River, the large thickets have been replaced by farms and pastures, the trees cut down, and the evergrowing population has crowded in on the old haunts of the cuckoos to such an extent that if they come here now at all they must be exceedingly rare."

In California, as throughout western North America, the last remaining groves of valley riparian forest are in jeopardy. Each year more of these forests are bulldozed and cut for pulpwood, or to make way for orchards, gravel extraction, rip-rap bank protection and urban development. Unless immediate measures are taken, this endangered habitat will no longer provide a home for the Yellow-billed Cuckoo and the many other birds and animals which dwell there.

As Eleanor Pugh (1965) recognized a decade ago, the choice is ours. "As long as housing tracts start landscaping from bare soil," she writes, "rather than plan around existing mature willows, cottonwoods, sycamores and oaks, with their entangled undergrowths so rich in the shyer birds; as long as willow shrub riparian cover is scraped away and replaced with ugly concrete channeling, breeding success will be low for many species. . . small wonder that Willow Flycatchers, Swainson's Thrushes, Yellowthroats, Yellow Warblers and Yellow-breasted Chats, though quite adaptive and once numerous, are becoming a rare sight to behold or even hear above the roar of traffic on the nearby freeway."

Acknowledgments

I wish to thank Sally Judy, Stephen Laymon and David Winkler for helpful suggestions and critical reading of the manuscript.

Literature Cited

American Ornithologist's Union 1957. Check-list of North American Birds. Fifth Edition.

Arbib, R. 1976. The blue list for 1977. Amer. Birds 30: 1031-1039.

Atwood, J. 1976. Winter bird population study. Disturbed floodplain woodland. Amer. Birds 30: 1061-1062.

Balda, R. P. 1975. Vegetational structure and breeding bird diversity. Proceedings of Symposium on Management Forest and Range Habitats for Nongame Birds. U.S.D.A. Forest Service technical report WO-1: 59-80.

Belding, L. 1878. A partial list of the birds of central California. Proc. U. S. Natl. Mus. 1: 388-449.

_____. 1890. Land birds of the Pacific district. Occ. Papers Calif. Acad. Sci. II.

Bendire, C. E. 1895. Life histories of North American birds. U. S. Nat. Mus. Spec. Bull. 3.

Bent, A. C. 1963. Life histories of North American birds. 20 volumes. Dover Publications, New York.

Best, L. B. 1975. Interpretational errors in the "mapping method" as a census technique. Auk 92: 452-460.

Chase, T. Jr. and P. DeBenedictis. 1965. Middle Pacific Coast region report. Aud. Field Notes 19: 71.

Cogswell, H. L. 1956. Middle Pacific Coast region report. Aud. Field Notes 10: 359.

_____. 1958. Middle Pacific Coast region report. Aud. Field Notes 12: 379-384.

Dawson, W. L. 1923. The birds of California. Four volumes. 2121 pp. South Moulton Co., San Diego.

Dembosz, D., K. Fickett and T. Manolis. 1972. Breeding bird census. Disturbed floodplain woodland. Amer. Birds 26: 978-979.

Desante, D. 1972. Breeding bird census. Riparian willow woodland. Amer. Birds 26: 1002-1003.

_____. 1976. The Changing seasons. American Birds 30: 677-689.

Emlen, J. T. 1938. Egrets of the Sacramento Valley. Gull 1: 41-42.

Enemar, A. 1959. On the determination of the size and composition of a passerine bird population during the breeding season-- a methodological study. Var Fagelvarld (Suppl.) 2: 1-104.

Friedmann, H. 1929. The cowbirds-- a study in the biology of social parasitism. 420 pp. C. C. Thomas, Springfield, Ill.

_____. 1963. Host relations of the parasitic cowbirds. U. S. Natl. Mus. Bull. 233.

Gaines, D. 1973. Breeding bird census. Floodplain riparian woodland. Amer. Birds 27: 995.

_____. 1974a. Review of the status of the Yellow-billed Cuckoo in California: Sacramento Valley populations. Condor 76: 204-209.

_____. 1974b. A new look at the nesting riparian avifauna of the Sacramento Valley, California. Western Birds 5: 61-80.

_____. ed. 1976. Abstracts from the Conference on the Riparian Forests of the Sacramento Valley. 25 pp. California Syllabus, Oakland, Calif.

Grinnell, J. 1915. A distributional list of the birds of California. Pac. Coast Avifauna 11.

Grinnell, J. and A. H. Miller. 1944. The distribution of the birds of California. Pac. Coast Avifauna 27.

Haartman, L. von. 1956. Territory in the Pied Flycatcher. Ibis 98: 461-475.

Hanna, W. C. 1928. Notes on the Dwarf Cowbird in Southern California. Condor 30: 161-162.

_____. 1937. California Cuckoo in the San Bernardino Valley, California. Condor 39: 57-59.

Hay, D. B. 1976. Winter bird population study. Freshwater marsh - riparian woodland. Amer. Birds 30: 1068.

Jewett, S. G. 1942. The European Starling in California. Condor 44: 79.

Ingles, L. 1950. Nesting birds of the willow-cottonwood community in California. Auk 67: 325-332.

Jay, A. 1911. Nesting of the California Cuckoo in Los Angeles County, California. Condor 13: 69-73.

Laskey, A. 1940. The 1939 nesting season of Bluebirds at Nashville, Tennessee. Wilson Bull. 52: 183-190.

MacArthur, R. H. and J. W. MacArthur. 1961. On bird species diversity. Ecology 42: 594-598.

Manolis, T. 1973. Breeding bird census. Disturbed floodplain woodland. Amer. Birds 27: 994-995.

Mayfield, H. 1965. The Brown-headed Cowbird, with old and new hosts. Living Bird 9: 13-28.

Metropulos, P. 1974. Breeding bird census. Disturbed floodplain woodland. Amer. Birds 28: 1035.

Miller, A. H. 1951. An analysis of the distribution of the birds of California. Univ. Calif. Publ. Zool. 50: 531-643.

Moffitt, J. 1939. Notes on the distribution of Sooty Shearwater, White Pelican and Cormorants. Condor 41:33.

71

Monson, G. 1960. Southwest region report. Aud. Field Notes 14: 469.

Muir, J. 1961. The mountains of California. 300 pp. Doubleday and Co., Garden City, New York.

Nice, M. M. 1937. Studies in the life history of the Song Sparrow I. A population study of the Song Sparrow. Trans. Linn. Soc. New York 4: 1-242.

Norris, R. T. 1947. The cowbirds of Preston Frith. Wilson Bull. 59: 83-103.

Ornduff, R. 1974. An introduction to California plant life. 152 pp. Univ. Calif. Press, Berkeley.

Pfeifer, S. 1963. Dichte und Dynamik von Brutpopulationen zweir deutscher Waldgebiete 1949-1961. Proc. XIIIth Internatl. Ornith. Congr. 2: 754-765.

Pugh, E. A. 1965. Southern Pacific Coast region report. Audubon Field Notes 19: 577.

Root, R. B. 1967. The niche exploitation pattern of the Blue-gray Gnatcatcher. Ecol. Monographs 37: 317-350.

Rowley, J. S. 1930. Observations on the Dwarf Cowbird. Condor 32: 130.

Salt, G. W. 1953. An ecologic analysis of three California avifaunas. Condor 55: 258-273.

Shannon, C. E. and W. Weaver. 1963. The mathematical theory of communication. 117 pp. Univ. of Ill. Press, Urbana.

Shuford, D. 1973. Breeding bird census. Disturbed riparian stream border. Amer. Birds 27: 1005.

Small, A. 1974. The birds of California. 310 pp. MacMillan Publishing Co., New York.

Southern, W. E. 1958. Nesting of the Red-eyed Vireo in the Douglas Lake region, Michigan. Jack-Pine Warbler 36: 105-130, 185-207.

Stewart, R. M. 1972. A summary of bird surveys in California. Pt. Reyes Bird Observatory Newsletter 21:3.

Storer, T. I. 1931. The American Egret in the lower Sacramento Valley, California. Condor 33: 34-35.

Svardson, G. 1957. The "invasion" type of bird migration. British Birds 50: 314-343.

Svensson, S. ed. 1970. Bird census work and environmental monitoring. Ecol. Res. Comm. Bull. 9.

Tangren, G. 1971. Breeding bird census. Riparian oak woodland. Amer. Birds 25: 967-969.

_____. 1972. Breeding bird census. Riparian oak woodland. Amer. Birds 26: 977-978.

Troetschler, R. G. 1976. Acorn Woodpecker breeding strategy as affected by Starling nest-hole competition. Condor 78: 151-165.

Vincenty, J. A. 1974. A study of factors affecting nesting raptor populations in urban areas, Sacramento County, California - 1974. Wildlife Management Branch Administrative Report No. 74-5. Calif. Dept. Fish and Game.

Welty, J. C. 1975. The life of birds. Second Edition. 623 pp. W. B. Saunders Co., Philadelphia.

Willet, W. C. 1933. A revised list of the birds of southwestern California. Pac. Coast Avifauna 21.

Willson, M. F. 1974. Avian community organization and habitat structure. Ecology 55: 1017-1029.

Winkler, D. 1973a. Winter bird population study. Riparian oak woodland. Amer. Birds 27: 689.

_____. 1973b. Breeding bird census. Riparian oak woodland. Amer. Birds 27: 996.

Table 1. The breeding riparian forest avifauna of the Sacramento Valley, California. See text for explanation of terms.

Species	status[a]	riparian affinity[b]	guild foraging	guild nesting	territorial males/km² x 2 cottonwood-willow 1A	1B	1C	2	3	oak-cottonwood 4A	4B	4C
Double-crested Cormorant (*Phalacrocorax auritus*)	res?	-	-	tree	-	-	-	-	-	-	-	-
Great Blue Heron (*Ardea herodias*)	res	-	-	tree	-	-	-	-	-	-	-	-
Green Heron (*Butorides virescens*)	mig	3	-	tree	-	-	-	-	-	+	-	+
Great Egret (*Casmerodius albus*)	res?	-	-	tree	-	-	-	-	-	-	-	-
Wood Duck (*Aix sponsa*)	res	2	-	tree hole*	-	-	-	-	+	-	+	-
Common Merganser (*Mergus merganser*)	res?	-	-	tree hole*	-	-	-	-	-	-	-	-
Turkey Vulture (*Cathartes aura*)	mig	8	ground carrion	tree stump	-	-	-	-	-	-	-	-
White-tailed Kite (*Elanus leucurus*)	res	1	ground mammal	tree	-	-	-	-	-	+	+	+
Cooper's Hawk (*Accipiter cooperi*)	res	1	foliage bird	tree	-	-	-	-	-	-	-	-
Red-tailed Hawk (*Buteo jamaicensis*)	res	5	ground mammal	tree	-	-	-	-	-	-	-	-

a res = resident; mig = migratory
b scale 1-8: 1 = primary affinity; 8 = species breeds in greater density in 7 other habitats (Miller 1951)
*does not excavate tree hole nesting cavity
**does excavate tree hole nesting cavity

Table 1, continued.

Species	status[a]	riparian affinity[b]	guild		territorial males/km^2 x 2							
					cottonwood-willow					oak-cottonwood		
			foraging	nesting	1A	1B	1C	2	3	4A	4B	4C
Red-shouldered Hawk *(Buteo lineatus)*	res	1	ground mammal	tree	+	+	+	-	-	+	-	+
Swainson's Hawk *(Buteo swainsoni)*	mig	-	-	tree	-	-	-	-	-	-	-	-
Bald Eagle *(Haliaeetus leucocephalus)*	res	-	-	tree	-	-	-	-	-	-	-	-
Osprey *(Pandion haliaetus)*	mig	-	-	tree	-	-	-	-	-	-	-	-
American Kestrel *(Falco sparverius)*	res	4	ground insect	tree hole*	-	-	-	-	-	-	-	-
California Quail *(Lophortyx californicus)*	res	-	ground seed	ground	80	86	52	60	94	156	170	138
Ring-necked Pheasant *(Phasianus colchicus)*	res	-	ground seed	ground	-	+	-	-	-	+	-	+
Mourning Dove *(Zenaida macroura)*	mig	3	ground seed	tree	70	52	52	+	-	+	+	+
Yellow-billed Cuckoo *(Coccyzus americanus)*	mig	1	foliage insect	tree	+	+	+	+	-	-	-	-
Screech Owl *(Otus asio)*	res	2	ground insect?	tree hole*	-	-	-	-	+	-	-	-
Great Horned Owl *(Bubo virginianus)*	res	4	ground mammal	tree	-	-	-	-	-	-	-	-
Long-eared Owl *(Asio otus)*	res?	1	ground mammal	tree	-	-	-	-	-	-	-	-
Anna's Hummingbird *(Calypte anna)*	res?	-	foliage nectar	tree	-	-	-	-	-	+	+	+
Black-chinned Hummingbird *(Archilochus alexandri)*	mig	1	foliage nectar	tree	-	-	-	-	-	-	-	-

Table 1, continued.

Species	status [a]	riparian affinity [b]	guild foraging	guild nesting	territorial males/km² x 2 cottonwood-willow 1A	1B	1C	2	3	oak-cottonwood 4A	4B	4C
Common Flicker (*Colaptes auratus*)	res	1	ground insect	tree hole**	+	+	+	+	+	+	+	+
Acorn Woodpecker (*Melanerpes formicivorus*)	res	-	foliage seed	tree hole**	-	-	-	-	94	56	78	70
Downy Woodpecker (*Picoides pubescens*)	res	1	bark insect	tree hole**	52	68	70	60	+	+	-	-
Nuttall's Woodpecker (*Picoides nuttalli*)	res	2	bark insect	tree hole**	70	86	88	100	94	+	+	+
Western Kingbird (*Tyrannus verticalis*)	mig	-	air insect	tree	52	+	+	-	-	-	-	-
Ash-throated Flycatcher (*Myiarchus cinerascens*)	mig	5	air insect	tree hole*	114	68	88	80	158	44	+	+
Black Phoebe (*Sayornis nigricans*)	res	2	air insect	-	+	-	+	-	+	-	+	-
Willow Flycatcher (*Empidonax traillii*)	mig	1	air insect	tree	-	-	-	-	-	-	-	-
Western Wood Pewee (*Contopus sordidulus*)	mig	4	air inect	tree	124	164	124	90	-	-	-	-
Tree Swallow (*Iridoprocne bicolor*)	mig	1	air insect	tree hole*	88	+	+	+	-	-	-	34
Purple Martin (*Progne subis*)	mig	-	air insect	tree hole*	-	-	-	-	-	-	-	-
Scrub Jay (*Aphelocoma coerulescens*)	res	-	generalist omnivore	tree	96	76	70	70	158	110	116	108
Yellow-billed Magpie (*Pica nuttalli*)	res	4	generalist omnivore	tree	-	-	-	-	-	-	+	-

Table 1, continued.

Species	status [a]	riparian affinity [b]	guild foraging	nesting	cottonwood-willow 1A	1B	1C	2	3	oak-cottonwood 4A	4B	4C
					territorial males/km² × 2							
Plain Titmouse (*Parus inornatus*)	res	-	bark insect	tree hole*	52	86	88	60	126	136	132	124
Bushtit (*Psaltriparus minimus*)	res	4	foliage insect	tree	52	52	34	60	+	100	84	92
White-breasted Nuthatch (*Sitta carolinensis*)	res	-	bark insect	tree hole*	-	-	-	-	-	+	+	+
Wrentit (*Chamaea fasciata*)	res	-	foliage insect	shrub	-	-	-	-	-	-	-	-
House Wren (*Troglodytes aedon*)	mig	2	foliage insect	tree hole*	-	-	-	60	-	-	+	46
Bewick's Wren (*Thryomanes bewickii*)	res	3	foliage insect	tree hole*	228	326	316	160	158	110	100	108
Mockingbird (*Mimus polyglottos*)	res	-	foliage insect	tree	-	-	-	-	-	+	+	+
California Thrasher (*Toxostoma redivivum*)	res	-	ground insect	shrub	-	-	-	-	-	-	-	-
American Robin (*Turdus migratorius*)	res?	6	ground insect	tree	52	68	70	100	-	+	+	+
Swainson's Thrush (*Catharus ustulata*)	mig	1	ground insect	tree	-	-	-	-	-	-	-	-
Blue-gray Gnatcatcher (*Polioptila caerulea*)	mig	4	foliage insect	tree	-	-	-	-	-	-	-	-
European Starling (*Sturnus vulgaris*)	res	-	generalist omnivore	tree hole*	52	52	52	-	252	204	200	200
Hutton's Vireo (*Vireo huttoni*)	res?	3	foliage insect	tree	-	+	-	-	-	-	-	-

77

Table 1, continued.

Species	status [a]	riparian affinity [b]	guild foraging	guild nesting	cottonwood-willow					oak-cottonwood		
					1A	1B	1C	2	3	4A	4B	4C
Bell's Vireo (*Vireo bellii*)	mig	1	foliage insect	tree	-	-	-	-	-	-	-	-
Warbling Vireo (*Vireo gilvus*)	mig	1	foliage insect	tree	-	-	-	-	-	-	-	+
Yellow Warbler (*Dendroica petechia*)	mig	1	foliage insect	tree	-	68	70	-	-	-	-	-
Common Yellowthroat (*Geothlypis trichas*)	mig	2	foliage insect	shrub	+	+	-	-	-	-	-	-
Yellow-breasted Chat (*Icteria virens*)	mig	1	foliage insect	shrub	+	+	-	-	-	-	-	-
House Sparrow (*Passer domesticus*)	res	-	ground seed	-	-	-	-	-	+	-	-	-
Western Meadowlark (*Sturnella neglecta*)	res	-	ground insect	ground	-	-	-	-	-	+	+	+
Northern Oriole (*Icterus galbula*)	mig	1	foliage insect	tree	88	120	194	100	-	-	-	-
Brown-headed Cowbird (*Molothrus ater*)	mig	1	ground seed	-	70	52	88	80	-	-	-	-
Black-headed Grosbeak (*Pheucticus melanocephalus*)	mig	1	foliage insect	tree	246	316	316	260	126	-	-	-
Blue Grosbeak (*Guiraca caerulea*)	mig	1	foliage insect	shrub	-	+	-	-	-	-	-	-
Lazuli Bunting (*Passerina amoena*)	mig	3	foliage insect	shrub	80	+	+	-	-	-	-	-
House Finch (*Carpodacus mexicanus*)	res	6	ground seed	tree	-	+	+	70	-	-	-	-

territorial males/km² × 2

78

Table 1, continued.

Species	status [a]	riparian affinity [b]	guild foraging	guild nesting	cottonwood-willow 1A	1B	1C	2	3	oak-cottonwood 4A	4B	4C
					territorial males/km² × 2							
American Goldfinch (*Carduelis tristis*)	res?	1	foliage seed	tree	-	52	70	160	-	-	-	-
Lesser Goldfinch (*Carduelis psaltria*)	res?	3	ground seed	tree	76	86	88	+	-	-	-	-
Rufous-sided Towhee (*Pipilo erythrophthalmus*)	res	2	ground seed insect	ground	184	188	212	120	126	124	116	92
Brown Towhee (*Pipilo fuscus*)	res	3	ground seed insect	shrub	88	+	34	+	94	+	+	+
Lark Sparrow (*Chondestes grammacus*)	res?	-	ground seed insect	ground	-	-	+	-	-	-	-	-
Song Sparrow (*Melospiza melodia*)	res	1	ground seed insect	shrub	-	-	-	-	94	-	-	-

CENSUS TOTALS

		1A	1B	1C	2	3	4A	4B	4C
Density	(males/km² × 2)	2032	2280	2268	1890	1968	1276	1250	1312
Species Richness	(# spp)	27	32	28	25	21	23	24	25
Species Diversity	($-\Sigma p_j \ln p_j$)	3.23	3.55	3.15	3.00	2.92	2.51	2.44	2.59

CENSUS SOURCES AND LOCATIONS

1A East bank of the Sacramento River north of the mouth of Chico Creek, Butte County — Dembosz, *et al.* 1972
1B same — Manolis 1973
1C same — Metropulos 1974
2 West bank of the Sacramento River 4.3 miles north of Glenn, Butte and Glenn Counties — Gaines 1973
3 Lake Solano County Park Campground 4.0 miles southwest of Winters, Solano County — Shuford 1973
4A Ancil Hoffmann Park, Carmichael, Sacramento County — Tangren 1971
4B same — Tangren 1972
4C same — Winkler 1973b

Table 2. The wintering riparian forest avifauna of the Sacramento Valley, California. res = resident; mig = migrant; spo = sporadic. C-W = cottonwood-willow; OAK = oak-cottonwood.

			Birds/km²			
			C-W		OAK	
Species	Status	Foraging Guild	1A	1B	2	3
Turkey Vulture (*Cathartes aura*)	res?	ground carrion	+	+	+	-
White-tailed Kite (*Elanus leucurus*)	res	ground mammal	-	-	+	+
Sharp-shinned Hawk (*Accipiter striatus*)	mig	foliage bird	+	-	+	+
Cooper's Hawk (*Accipiter cooperii*)	mig res?	foliage bird	-	-	-	+
Red-tailed Hawk (*Buteo jamaicensis*)	res mig	ground mammal	+	+	29	+
Red-shouldered Hawk (*Buteo lineatus*)	res	ground mammal	-	-	-	+
American Kestrel (*Falco sparverius*)	res	ground insect	-	-	+	+
California Quail (*Lophortyx californicus*)	res	ground seed	74	148	62	147
Band-tailed Pigeon (*Columba fasciata*)	spo	foliage seed	-	-	-	170
Mourning Dove (*Zenaida macroura*)	res	ground seed	-	-	+	+
Barn Owl (*Tyto alba*)	res?	ground mammal	-	+	+	-
Screech Owl (*Otus asio*)	res	ground insect?	-	-	-	-
Great Horned Owl (*Bubo virginianus*)	res	ground mammal	+	-	+	-

Table 2, continued.

Species	Status	Foraging Guild	C-W		OAK	
			1A	1B	2	3
Long-eared Owl (*Asio otus*)	mig?	ground	-	-	-	-
Anna's Hummingbird (*Calypte anna*)	res?	bark insect	10	10	-	39
Common Flicker (*Colaptes auratus*)	res mig	ground insect; foliage seed	20	20	54	131
Acorn Woodpecker (*Melanerpes formicivorus*)	res	foliage seed	+	+	50	185
Yellow-bellied Sapsucker (*Sphyrapicus varius*)	mig	bark sap	+	+	+	+
Downy Woodpecker (*Picoides pubescens*)	res	bark insect	20	15	17	+
Nuttall's Woodpecker (*Picoides nuttalli*)	res	bark insect	27	25	25	31
Black Phoebe (*Sayornis nigricans*)	res	air insect	-	10	+	-
Scrub Jay (*Aphelocoma coerulescens*)	res	generalist omnivore	27	44	141	479
Yellow-billed Magpie (*Pica nuttalli*)	res	generalist omnivore	-	-	72	-
Common Crow (*Corvus brachyrhynchos*)	res	generalist omnivore	+	+	21	-
Plain Titmouse (*Parus inornatus*)	res	bark insect; foliage seed	17	20	66	62
Bushtit (*Psaltriparus minimus*)	res	bark insect	15	158	66	62

Table 2, continued.

| | | | Birds/km² | | | |
| | | | C-W | | OAK | |
Species	Status	Foraging Guild	1A	1B	2	3
White-breasted Nuthatch (*Sitta carolinensis*)	res	bark insect	10	25	25	15
Brown Creeper (*Certhis familiaris*)	spo	bark insect	+	25	+	-
Wrentit (*Chamaea fasciata*)	res	bark insect	-	-	+	-
Winter Wren (*Troglodytes troglodytes*)	mig	bark insect	+	+	-	-
Bewick's Wren (*Thryomanes bewickii*)	res?	bark insect	35	30	12	39
Mockingbird (*Mimus polyglottos*)	res	bark insect	-	+	+	+
American Robin (*Turdus migratorius*)	res mig	foliage seed; ground insect	124	10	-	85
Varied Thrush (*Ixoreus naevius*)	spo	foliage seed; ground insect	-	-	-	108
Hermit Thrush (*Hylocichla guttata*)	mig	ground insect: foliage seed	17	15	17	+
Western Bluebird (*Sialia mexicana*)	mig?	foliage seed: air insect	-	-	37	-
Golden-crowned Kinglet (*Regulus satrapa*)	spo	bark insect	-	20	12	-
Ruby-crowned Kinglet (*Regulus calendula*)	mig	bark insect; air insect	69	99	29	46
Cedar Waxwing (*Bombycilla cedrorum*)	mig	foliage seed	-	+	108	39

Table 2, continued.

Species	Status	Foraging Guild	Birds/km²			
			C-W		OAK	
			1A	1B	2	3
Phainopepla *(Phainopepla nitens)*	res?	foliage seed; air insect	-	-	+	-
European Starling *(Sturnus vulgaris)*	res?	generalist omnivore	27	32	108	432
Hutton's Vireo *(Vireo huttoni)*	mig?	bark insect	-	+	-	-
Orange-crowned Warbler *(Vermivora celata)*	mig	bark insect	10	17	+	-
Audubon's Warbler *(Dendroica coronata auduboni)*	mig	bark insect; air insect	94	49	25	100
Myrtle Warbler *(Dendroica coronata hooveri)*	mig	bark insect; air insect	+	-	+	15
Western Meadowlark *(Sturnella neglecta)*	res	ground insect	-	-	-	15
Evening Grosbeak *(Hesperiphora vespertina)*	spo	foliage seed	+	-	-	-
Purple Finch *(Carpodacus purpureus)*	spo	foliage seed	22	+	-	+
House Finch *(Carpodacus mexicanus)*	res	ground seed	-	-	-	100
Pine Siskin *(Carduelis spinus)*	spo	foliage seed	-	-	+	+
American Goldfinch *(Carduelis tristis)*	res?	foliage seed	44	65	-	15
Lesser Goldfinch *(Carduelis psaltria)*	res	ground seed	-	32	46	131

Table 2, continued.

| Species | Foraging Guild | Status | Birds/km² | | | |
| | | | C-W | | OAK | |
			1A	1B	2	3
Rufous-sided Towhee (*Pipilo erythrophthalmus*)	ground seed	res mig	25	25	71	131
Brown Towhee (*Pipilo fuscus*)	ground seed	res	15	42	33	-
Dark-eyed Junco (*Junco hyemalis*)	ground seed	mig	35	37	71	108
White-crowned Sparrow (*Zonotrichia leucophrys*)	ground seed	mig	22	30	199	108
Golden-crowned Sparrow (*Zonotrichia atricapilla*)	ground seed	mig	+	+	324	208
White-throated Sparrow (*Zonotrichia albicollis*)	ground seed	mig	+	-	+	+
Fox Sparrow (*Passerella iliaca*)	ground seed	mig	+	+	+	-
Lincoln's Sparrow (*Melospiza lincolnii*)	ground seed	mig	25	+	-	-
Song Sparrow (*Melospiza melodia*)	ground seed	mig res	49	27	-	-
CENSUS TOTALS						
Density (birds/km²)			880	1114	1720	3158
Species Richness (# species)			38	41	46	41
Species Diversity ($- p_i \ln p_i$)			2.92	3.00	3.00	3.11

CENSUS SOURCES AND LOCATIONS

1A Dog Island, Red Bluff, Tehama County   Stephen Laymon, unpub. data, 1975
1B same   "   "   "   1976
2 Bobelaine Audubon Sanctuary, 11 miles SSE of Yuba City, Sutter County   Atwood 1975
3 Ancil Hoffmann Park, Carmichael, Sacramento County   Winkler 1973a

Table 3. Peak numbers (birds/km^2 of selected migrants in cottonwood-willow riparian forest. Data from Dog Island, Red Bluff, Tehama County (Stephen Laymon, Unpublished censuses).

| Species | Spring Peak | | | | Fall Peak | |
| | 1975 | | 1976 | | 1976 | |
	Date	#	Date	#	Date	#
Ash-throated Flycatcher (*Myiarchus cinerascens*)	5-14	20	5-9	20	7-12	30
Willow Flycatcher (*Empidonax traili*)	4-25	5	5-5	5	8-15	59
Western Flycatcher (*Empidonax difficilis*)	-	-	5-5	5	8-15	10
Solitary Vireo (*Vireo solitarius*)	-	-	4-27	10	-	-
Warbling Vireo (*Vireo gilvus*)	5-6	25	5-3	35	9-7	5
Orange-crowned Warbler (*Vermivora celata*)	4-17	123	4-11	104	9-20	79
Nashville Warbler (*Vermivora ruficapilla*)	4-25	64	5-2	10	8-1	30
Yellow Warbler (*Dendroica petechia*)	5-21	79	5-19	35	9-14	193
Black-throated Gray Warbler (*D. Nigrescens*)	5-8	10	4-27	5	10-6	25
Townsend's Warbler (*Dendroica townsendi*)	5-8	15	5-7	15	-	-
Hermit Warbler (*Dendroica occidentalis*)	-	-	5-19	5	-	-
MacGillivray's Warbler (*Oporornis tolmiei*)	5-3	10	4-27	20	8-27	44
Common Yellowthroat (*Geothylpis trichas*)	4-25	35	4-11	44	9-20	25
Wilson's Warbler (*Wilsonia pusilla*)	5-4	104	5-3	207	9-14	25
Northern Oriole (*Icterus galbula*)	?	?	?	?	7-12	163
Western Tanager (*Piranga ludoviciana*)	5-14	109	5-13	35	8-1	59
Lazuli Bunting (*Passerina amoena*)	5-21	40	5-3	5	-	-

Chapter 8

HABITATS OF NATIVE FISHES IN THE SACRAMENTO RIVER BASIN

Donald W. Alley, D. H. Dettman, H. W. Li, and P. B. Moyle
Wildlife and Fisheries Biology
University of California, Davis

Introduction

It is a curious fact that many of the world's finest ichthyologists have resided in California: David Starr Jordan, Carl Hubbs, Earl Herald, George Meyers, Warren Freihoffer, and Robert Behnke, yet there is little understanding of the ecology of native freshwater fishes of the Sacramento drainage. Perhaps this can be attributed to the fact that most of the work in California was taxonomic, rather than ecologically oriented. In addition, most of the ecological studies were done either on the eastern slope of the Sierras by the late Paul R. Needham and his students or on game species by the California Department of Fish and Game. In fact, nongame fishes were considered to compete or interfere with game fishes and programs of eradication instituted as a cure for declining game fish production. These programs were not based on careful analysis nor ecological studies. However, in recent years, considerable interest in the native nongame fishes has developed and a number of research programs concerning their ecology and status are now well underway, particularly at UC Davis. Most of this work to 1975 is reviewed in Moyle and Koch (1975) and Moyle (1976b). The purpose of this paper is to bring these reviews up to date by summarizing our progress in trying to understand the ecology of California's native fish communities, particularly in the Sacramento Drainage Basin, through surveys and experiments.

Habitat Changes

The task of understanding the stream fish communities has been difficult because the streams of California have been badly disturbed. The destruction of the riparian forests has been only one part of this perturbation, although one of the most visible. One of the first major disturbances was placer mining which destroyed salmonid spawning grounds, increased siltation, removed or covered up riparian vegetation, and drastically changed stream morphology. As agriculture became more and more important to California's economy the deterioration of aquatic habitats continued (and continues) at an ever-increasing rate. The first major alterations caused by agriculture were the draining of tule beds and wetlands on the valley floor. These areas probably acted as nursery grounds for many native fishes and their destruction may have been the primary cause for the disappearance of Gila crassicauda, the thicktailed chub. Then, as irrigation and flood control became necessary, channelization of streams started to become common as did irrigation diversions and the construction of bypasses for flood waters. Channelization consists of vegetation removal, straightening channels (thus removing meanders), dredging the stream bed and stabilizing the banks with loose material (riprapping). This type of habitat alteration has been well documented in terms of its effect (Whitney and Bailey, 1959; Peters and Alvord 1964; Funk and Ruhr, 1971; Barton, et al., 1972; Moyle 1976a). Essentially, the environment has been eliminated, and undercut banks are destroyed. The substrate is made more uniform as snags and fallen logs are removed. As expected, species richness and standing crops diminish as a result. Irrigation diversions and flood bypasses often divert migratory young of anadromous fishes from the main streams. The degree of impact of these diversions is not presently known; however, there is some concern that substantial mortality of young may contribute to declining chinook salmon and steelhead runs. Dewatering streams for irrigation also reduces flows, which triggers a series of changes: water temperatures increase, current is reduced, silt deposition increases, dissolved oxygen decreases, the stream becomes more shallow, and finally production decreases. Perhaps the worst treated of streams used as sources of irrigation water have been intermittant streams and small tributaries that are often allowed to be dewatered completely. The reason for this is that it is generally felt that they are unimportant for fish production. However, recent work by Erman and Hawthorne (1976) suggests that this concept should be reevaluated. They found that 39-47% of the spawning population of rainbow trout in Sagehen Creek spawned in an intermittant tributary to the Creek. The key point here is that the stream did not become dry until the young fish had hatched and moved out, before mid summer. In the Central Valley, such streams appear to be important to chinook salmon, as well as to native fishes such as Sacramento squawfish and Sacramento suckers. All three species still spawn in lower Putah Creek when there is enough rain for it to flow in the winter. The young salmon move out completely in the spring, while the young squawfish and suckers will stay in the deep pools that usually remain in a few places throughout the summer.

Grazing is another activity which has altered fish habitat, particularly on small streams. This is noticeable around areas like Chico, Orland, and Vina. Grazing causes banks to collapse because of trampling and removes overhanging grasses and bushes. As a result, cover for fish is greatly reduced. Silt becomes deposited in the stream due to increased local erosion. This reduces production of aquatic insects and the number of suitable spawning sites for many fishes, but especially trout (White and Brynildson, 1967).

Bad forestry practices can lead to severe problems very similar to overgrazing of livestock. Small streams are often used as chutes to transport downed trees, badly damaging banks and substrate. If slash is dumped into creeks, this will cause dams to form which will impede spawning migrations, decrease flow, and increase siltation. Major problems in logging areas have also been caused by poorly designed roads which often follow stream courses. Such roads can accelerate soil erosion tremendously which dramatically increases the silt burden of the stream (Platts and Megahan, 1975; Megahan and Kidd, 1972; Arnold and Lundeen, 1968). Fine sediment can smother embryos, alevins, and fry. Fish migrations may also be impeded when roads cross the stream and improperly designed conduits are constructed. Conduits that create waterfalls or accelerate flow excessively are often barriers to fish movements. It is worth noting here that proper forestry management practices can actually increase salmon and trout production (Burns, 1972) as it is well known that too much shade will inhibit food production (White and Brynildson, 1967).

The most visible change in the Sacramento drainage has been the construction of dams for water storage, hydroelectricity, flood control and groundwater recharge. These dams have had an effect on the fish fauna both upstream and downstream from the dams. Upstream the dams block access of migratory salmonids to their spawning grounds and create totally new habitats with the filling of the reservoirs. The upstream spawning grounds may be lost even if fish ladders are constructed because the migrating fish often become lost in the new reservoirs. A similar problem exists for juvenile fish moving downstream even if the parent fish had been able to find the spawning grounds. In addition, large losses of juvenile fish may occur if they have to pass through the turbines of hydroelectric plants. Although a number of mitigation hatcheries have been constructed in California, in no case have these compensated for initial losses of salmonids dependent on upstream spawning grounds. The reservoirs are essentially lakes, so it is not surprising to find that the native stream dwelling fishes often are unable to survive in them. Although native bottom feeders such as suckers, blackfish, and sculpins may consequently increase in the reservoir, reservoirs, in general, favor exotic over native fishes as the native fishes have evolved for living in rivers. These changes can also alter the distribution of fishes upstream from the reservoir. Erman (1973) found a dramatic change in fish distribution in Sagehen Creek following the downstream impoundment of Stampede reservoir. Suckers (Catostomus tahoensis) and whitefish (Prosopium williamsoni) are now found at higher elevations than previously recorded. There are several ways that habitat downstream from dams can be altered. (1) Downstream temperature can be changed. For example, summer temperatures in the Sacramento River below Shasta Dam are considerably lower than they used to be because of the rebase of cold water from the reservoir. As a consequence, salmon and trout are now dominant where native warm water fishes used to live. (2) Daily fluctuations of water release may disturb natural cycles of feeding behavior and aquatic insect drift. This may profoundly alter microhabitat selection and thus cause unstable social hierarchies, through fluctuations in territory size. Although this has not been tested, we suspect this may be true because stream velocity is the primary factor influencing fish distribution (Dettman and Li in preparation, Alley and Li in press). (3) The more constant year around flow of the dammed stream may alter fish communities through the reduction of winter washouts and the scouring effects of winter flooding. The winter floods are important in renewing gravels and creating more habitat by such means as deepening pools and causing trees to fall into the water. (4) The reservoir may act as a nutrient trap.

Dams have also had an indirect negative impact upon native fish communities because they encourage fish poisoning programs. Often the first step taken to manage a new reservoir is to poison out the "trash fish" (i.e., native fishes) both in the reservoir area and upstream from it so that they will not compete or prey upon game fishes. These operations are rarely completely effective, so the native species are often able to reestablish themselves in large numbers. This increase may coincide with the normal decline in harvest of game fishes probably at least partially caused by over fishing. Interspecific competition is invoked to explain the game fish decline and another poisoning operation is done to "rejuvenate" the fishery. Another indirect effect of dams is that some of the introduced fishes escape downstream and alter food webs and community structure in downstream areas. It is worth noting, however, that the exotic fishes usually manage to become established only in areas where the habitat has been altered.

This leads into another important topic, the impact of exotic fishes on native fish communities which will not be discussed any further here since it is covered in detail in Moyle (1976b, c). We would, however, like to leave you with this one thought in this area. What impact do striped bass have on downstream migrants of steelhead and chinook salmon? We know that the bass prey on them, but not the magnitude of predatory losses.

Fish Communities

The assemblege of native freshwater fishes in the Sacramento drainage is typified by a relatively large number of endemic species (17) and a large number of monotypic genera (Table 1) contains minnows (Cyprinidae), trout and char (Salmonidae), sculpins (Cottidae), suckers (Catostomidae), one centrarchid (Sacramento perch, Archoplites interruptus), and a surfperch which has evolved a freshwater habitus (tule perch, Hysterocarpus traskii). The obligatory freshwater fauna is augmented by several families of anadromous fishes: lamprey (Retromyzontidae), sturgeon (Acipenseridae), salmon (Salmonidae), and sticklebacks (Gasterosteidae). In total there are 55 native species and 36 exotic species in the Sacramento drainage.

Faunal elements are shared between the Sacramento and San Joaquin River system, but the Sacramento system is richer and more complex. This is because the discharge is greater and the drainage basin is larger and thus tributaries of the Sacramento collect runoff from a wider variety of habitats. Because of the lack of discharge, the San Joaquin has steeper changes in velocity and temperature gradients. Zones of fish assembleges are very distinct. For instance, Moyle and Nichols (1973, 1974) were able to recognize four major fish associations which were aligned in order of elevation: (1) introduced fish associations, (2) native cyprinid-catostomid association, (3) California roach association, and (4) the rainbow trout association. The introduced fishes were primarily warm water game fishes and were found in low gradient, disturbed environments (mostly lacking riparian forests). Fishes found in the foothills of the Sierra were the hardhead (Mylopharodon conocephalus), squawfish (Ptychocheilus grandis), and Sacramento sucker (Catostomus occidentalis) which inhabited warmer pools and slow deep runs in undisturbed streams. The California roach (Hesperoleucas symmetricus) were found in small, clear, intermittent streams, but these fishes seemed to be particularly well adapted for dessicating pools, tolerating low levels of dissolved oxygen. The rainbow trout association was comprised of a few suckers and an abundance of rainbow trout. These associations are similar to those of Murphy (1948) and Hopkirk (1967) for sections of the Sacramento, but because of human disturbance and less permanence of the streams of the San Joaquin, which are all eastward draining, there is less species richness and diversity. Once the San Joaquin leaves the foothills it becomes a warm, sluggish, turbid stream.

Zonation is not as sharp and clear cut in the Sacramento River system. As it collects waters from many tributaries, it has a wide variety of temperature regimes in different sections of the river, depending on whether the inflowing tributary is cold (McCloud River) or warm (Deer Creek). The discharge is so much greater than the San Joaquin that one can find cold water species in the main river itself. Thus one can observe an abstracting process in a tributary such as Deer Creek, where the zones are very similar to that of the tributaries of the San Joaquin (Dettman and Li ms, Alley and Li in press). However, at the confluence of Deer Creek and the Sacramento River cold water fauna is found again. This pattern is particularly pronounced in the Pit River system where one finds interspersed, cold water fauna and warm water fauna in different stretches along the stream depending on (1) water velocity, or local gradient, (2) stream order, (3) average stream depth, and (4) temperature regime (Moyle and Daniels ms). The latter factor is greatly influenced by the nature of the proximate inflowing tributary as well as by the development of the riparian habitat. Well-shaded streams are generally much cooler than those that are not.

In order to see how these factors affect the native fishes, we have extensively studied fish behavior in Deer Creek, California near Chico. This stream was selected for study because it is a minimally disturbed system. There is some grazing near Vina, a small diversion dam, and smallmouth bass and hatchery rainbow trout have been planted.

89

Table 1. Distributional checklist of fishes in the Sacramento drainage.
N = native, I = introduced, O = occasional, A = anadromous,
E = extinct, R = rare, ? = status uncertain.

Species	Status	Species	Status
Petromyzontidae		Ictaluridae	
Pacific lamprey	NA	Channel catfish	I
River lamprey	NA	White catfish	I
Pacific brook lamprey	N	Yellow bullhead	I
Pit-Klamath brook lamprey	N	Brown bullhead	I
		Black bullhead	I
Acipenseridae			
White sturgeon	NA	Cyprinodontidae	
Green sturgeon	NA	Rainwater killifish	I
Clupeidae		Poeciliidae	
Pacific herring	O	Mosquitofish	I
American shad	IA		
Threadfin shad	I	Atherinidae	
		Topsmelt	O
Osmeridae		Mississippi silverside	I
Delta smelt	N/I		
Surf smelt	O	Gasterosteidae	
Longfin smelt	N	Threespine stickleback	NA
Salmonidae		Syngnathidae	
Pink salmon	NA	Bay pipefish	O
Chum salmon	NA		
Coho salmon	NA/I	Percichthyidae	
Chinook salmon	NA	Striped bass	I
Sockeye salmon	NA		
Kokanee	I	Centrarchidae	
Brook trout	I	Sacramento perch	NR
Interior Dolly Varden	NR	Black crappie	I
Coast Dolly Varden	I	White crappie	I
Brown trout	I	Warmouth	I
Redband trout	NR	Green sunfish	I
Golden trout	I	Bluegill	I
Rainbow trout	N	Pumpkinseed	I
Arctic grayling	I	Redear sunfish	I
		Largemouth bass	I
Cyprinidae		Spotted bass	I
Carp	I	Redeye bass	I
Goldfish	I		
Tench	I	Percidae	
Golden shiner	I	Yellow perch	IE
Sacramento blackfish	N	Bigscale logperch	I
Hardhead	N		
Hitch	N	Embrotocidae	
Sacramento squawfish	N	Shiner perch	O
Tui chub	NE/I	Tule perch	N
Thicktail chub	NE		
Sacramento splittail	N	Gobiidae	
California roach	N	Tidewater goby	N
Speckled dace	N	Yellowfin goby	I
Lahontan redside	I	Arrow goby	O
Red shiner	I	Chameleon goby	?
Fathead minnow	I		
		Cottidae	
Catostomidae		Staghorn sculpin	N
Mountain sucker	N?	Coastrange sculpin	N
Modoc sucker	NR	Prickly sculpin	N
Tahoe sucker	I?	Riffle sculpin	N
Sacramento sucker	N		
		Pleuronectidae	
		Starry flounder	N

90

None of these fishes have been found within Deer Creek Canyon, the area of our study sites. Public access is restricted in the middle reaches of Deer Creek and most of the northern watershed is within the confines of a state wildlife refuge. It originates in the Sierra Nevada Mountains of Tehama County at an elevation of 1550 m and flows for approximately 118 km until its confluence with the Sacramento River near Vina, California (elevation, 63 m). By studying Deer Creek, we believe that we could gain some understanding of part of the Sacramento system and thus determine the impact of human activity on other sections of the basin. We have found that (1) stream velocity, (2) maximum daily temperature, and (3) percentage of pools were the factors which had the greatest influence on fish distribution. As one surveys Deer Creek beginning from the Canyon one finds flows of 18-19 cm/sec, water temperatures of 20.5-23.5°C, and the bottom is silty; this stretch has few riffles, being composed primarily of deep runs and pools (Dettman and Li ms). Large squawfish, hardhead, suckers, roach and tule perch and an occasional smallmouth bass are found here. As one proceeds up the Canyon, average stream flow becomes faster (28.9 cm/sec), although the temperature does not change; a third of the section may be composed of riffles, 60% runs and less than 10% pools. Tule perch and smallmouth bass drop out, but trout and sculpin are found in the riffles and the large hardhead squawfish are found in the runs and pools. Suckers are eclectic in habitats occupied. They feed diurnally and are foraging in the fast sections in the middle of the river. Speckled dace, small minnows, are found in the interstices of gravel and cobble nearer the banks. Roach are found in the quiet backwaters of pools. Further upstream, the average water velocity increases to 40-50 cm/sec, the water velocity becomes cooler (16-20°C) and the number of riffles increases. Squawfish drop out; hardhead are found in deeper runs. Juvenile salmon are either inshore under cover and in pools with the roach or in the faster sections of the stream, depending upon the presence or absence of squawfish, which prey upon them. Trout are now found in the head of deep runs and pools as well as riffles. Nearer the headwaters, the cyprinids (minnows) drop out except for an occasional speckled dace. Suckers are found in fewer numbers, but not uncommon, and rainbow trout and sculpins are found in abundance. Average stream velocity becomes extremely swift, 50-70 cm/sec, and the water is cold, 13-18°C. Riffles, runs and pools now occur in near equal frequency. The size of the rocky substrate increases as one heads upstream. In comparison with the tributaries of the San Joaquin, the transitions were not abrupt; there was a more gradual abstraction of species with elevation.

In order to assess the role of physical factors relative to interspecific competition, we have been measuring species densities, microhabitat preferences, metabolic performance and swimming ability. Transects were set up, population estimates made, locations of fishes were mapped and site specific parameters recorded, including stream velocities at focal points of fish activities. Measurements were then taken of tail beat frequencies at different times of the day and feeding behavior to get some measure of how fishes partitioned time and activity. Swimming stamina and metabolic rate measurements of hardhead minnows were made in Brett-type swimming respirometers (Brett 1964). Individual fish were trained to swim against a current in order to avoid shock from an electrical field at the end of the respirometer. Frequency of tail beats, oxygen uptake and swimming velocity were correlated. The model provided by Feldmeth and Jenkins (1973) was used to measure tail beat frequency and metabolic performance for rainbow trout.

Microhabitat studies reveal that rainbow trout selected significantly faster water than suckers, hardhead minnows, squawfish and roach. There were differences in velocity selection by trout in two different sites because of different gradients in each section, but more than 50% of the trout inhabited currents with velocities between 30 and 70 cm/sec and 20% of those measured were found swimming at velocities greater than 70 cm/sec (n = 172). More than 60% of the suckers selected velocities between 10 and 50 cm/sec with 15% of them in currents greater than 70 cm/sec (n = 179). Suckers in relation to trout stayed on the stream bottom and used little effort in maintaining position, presumably because the hydrodynamic, "spoiler effect" of the conformation of the sucker allows it to hug the bottom. Most of the cyprinids were recorded in water velocities less than 30 cm/sec (n = 373).

Differences in velocity and depth preferences resulted in spatial segregation of some of the species in upper and lower canyon study sites. In the upper canyon areas, trout generally inhabited riffles and swift heads of pools as well as its shallow tails. Suckers inhabited similar areas to the trout but were beneath the trout when both species were together.

In the upper canyon the squawfish and hardhead generally inhabited slow, deep areas of pools. Hardhead minnows were found occasionally in slower, deeper portions of runs where some trout maintained positions during the day. No behavioral interactions between squawfish and trout were observed.

In the warmer, lower canyon areas species distributions were different. Trout were restricted to the riffles in August and September, 1975. No trout were seen in the pools until the middle of November when water temperatures were greatly cooler. However, hardhead and squawfish were widely distributed in the lower canyon in August and September rather than the upper canyon. For example, hardhead were not only seen in deep, slow pools but also in the heads and tails of pools, in runs and slower portions of riffles. Squawfish and hardhead greater than 20 cm in length behave differently. Squawfish do not hold a position in the current but spend their time cruising around. Feeding is nocturnal. Large and small hardhead will hold positions in the currents to feed on drifting insects, although they are primarily midwater and bottom pickers. They feed diurnally. Trout and salmon are surface drift feeders, feeding most actively during crepuscular periods. We observed no intraspecific territoriality between hardhead during feeding, though, as was observed among trout.

Hardhead and squawfish were considerably more abundant in the lower canyon areas rather than in the cooler, upper canyon. The reverse was true for trout and salmon, although transects through 19 study sections demonstrated that patterns of trout abundance varied independently of squawfish density (Dettman and Li ms). Suckers were more abundant downstream.

Laboratory studies revealed that trout have higher metabolic rates and higher metabolic scope for activity at the same temperatures than hardhead minnows (Alley and Li in press). But rainbow trout were far more powerful and efficient swimmers than hardhead. Thus a hardhead minnow will consume less oxygen at 20°C than a rainbow trout of a similar weight acclimated to 15°C. We also found in simplified communities in laboratory streams that trout were far more aggressive than cyprinids and suckers and that the dominant trout determined where other fishes were stationed through aggressive behavior.

In conclusion, the results of these studies strongly indicate that changes in species composition below dams are not caused by competition between species for space. Patterns of abundance seemed to be determined by physical factors which affect physiological responses of fishes which in turn may influence behavior. Trout are well adapted for fast, cool water and feeding in sections of high velocities, such as riffles and heads of pools where maximal amounts of insect drift occur. Trout have high food requirements because of their high metabolic rate. Occupying warmer areas increase the cost/benefit ratios of inhabiting fast sections, but restrict trout to these areas because slower areas have less drift per unit time and bottom foraging is calorically expensive. Thus trout are not found in pools in warmer areas. Since shade from riparian forests help keep streams cooler, it seems likely that removal of the forests will contribute to the decline of salmonid populations.

Cyprinids are not as efficient as nor can they sustain as high swimming speeds as trout. In cooler areas, their metabolic scope for performance is limited, thus they can maintain themselves only in pools. In warmer sections their metabolic scope increases and they are more efficient than trout, thus they can inhabit faster runs although they are mainly found in pools. Because of their lower metabolic requirements, they can attain greater densities than trout in warmer sections.

We suggest that on the basis of water velocity and temperature that the sympatric zone in Deer Creek is marginal habitat for both trout and cyprinids. Displacement of trout by cyprinids is unlikely to be caused by behavioral aggression. None were observed in Deer Creek and laboratory studies indicate that trout are aggressively dominant in any case.

Numerous water diversion projects completed in California during the past 60 years have drastically altered natural hydrologic factors and increased water temperatures. A 90% reduction in flow caused average width, depth, and velocity to decrease by 22%, 44%, and 75% (Curtis 1959). A similar reduction in flow can result in a 75% decrease in riffle area, a 55% increase in shallow runs, and 96% decrease in deep, fast runs (Kraft 1972). This type of disruption of the natural hydrologic regime can explain recent imbalances in native fish populations and is more probable than interspecific competition.

The destruction of riparian forests in the Central Valley has been a small, but probably important, factor contributing to the changes in the fish communities, mostly because of the effect on water temperature. However, there is much we do not understand about their relationship to fish populations, particularly in regard to the use of flooded vegetation by young fish and the role of logs and other debris in increasing habitat diversity.

Literature Cited

Alley, D. W. and H. W. Li. In press. Significance of microhabitat selection for fishes in a Sierra foothill stream. Cal-Neva Trans. Amer. Fish. Soc.

Arnold, J. F. and L. Lundeen. 1968. South fork of the Salmon River special survey - soils and hydrology. USDA For. Serv., Intermountain Region. Mimeographed report.

Barton, J. R., E. J. Peters, D. A. White and P. V. Winger. 1972. Bibliography on the physical alteration of the aquatic habitat (Channelization) and stream improvement. Brigham Young Univ. Publ., Provo, Utah. 30 pp.

Brett, J. R. 1964. The respiratory metabolism and swimming performance of young sockeye salmon. J. Fish. Res. Board Can. 20:1183-1226.

Burns, J. W. 1972. Some effects of logging and associated road construction on Northern California streams. Trans. Amer. Fish. Soc. 101(1): 1-17.

Curtis, B. 1959. Changes in a river's physical characteristics under substantial reduction in flow due to hydroelectric diversion. Calif. Fish Game 45:181-188.

Dettman, D. H. and H. W. Li. Ms. Macrohabitat factors influencing the distribution and abundance of rainbow trout (Salmo gairdneri) and Sacramento squawfish (Ptychocheilus grandis) in Deer Creek, California.

Erman, D. C. 1973. Upstream changes in fish populations following impoundment of Sagehen Creek, California. Trans. Amer. Fish Soc. 103:626-629.

_____ and V. M. Hawthorne. 1976. The quantitative importance of an intermittent stream in the spawning of rainbow trout. Trans. Amer. Fish Soc. 105:675-681.

Feldmeth, C. R. and T. M. Jenkins. 1973. An estimate of energy expenditure by Rainbow Trout (Salmo giardneri) in a small mountain stream. J. Fish Res. Bd. Can. 30: 1755-1759.

Funk, J. L. and C. E. Ruhr. 1971. Stream channelization in the midwest. p. 5-11. In E. Schneberger and J. L. Funk, eds. Stream channelization: a symposium. N. Cent. Div. Amer. Fish. Soc. Spec. Publ. 2.

Hopkirk, J. D. 1967. Endemism in fishes of the Clear Lake region. Ph.D. thesis. Univ. Calif., Berkeley.

Kraft, M. E. 1972. Effects of controlled flow reduction on a trout stream. J. Fish. Res. Board Can. 29:1405-1411.

Li, H. W. 1975. Competition and coexistence in stream fishes. p. 19-30. In P. B. Moyle and D. Koch, eds. Symposium on trout/nongame fish relationships in streams. Des. Res. Inst. Wat. Res. Center Rpt. No. 17.

Megahan, W. F. and W. J. Kidd. 1972. Effects of loggina and logging roads on erosion and sediment deposition from steep terrain. J. For. 80:136-141.

Moyle, P. B. and D. L. Koch. 1975. eds. Symposium on Trout/Non-game Fish Relationships in Streams. Center for Water Resources Research. Desert Research Institute Misc. Report 17. 81 pp.

Moyle, P. B. and R. D. Nichols. 1973. Ecology of some native and introduced fishes of the Sierra Nevada foothills in Central California. Copeia (3):478-490.

Moyle, P. B. and R. D. Nichols. 1974. Decline of the native fish fauna of the Sierra Nevada foothills, Central California. Amer. Midl. Natur. 92(1):72-83.

_____. 1976a. Some effects of channelization on the fishes and invertebrates of Rush Creek, Modoc County, California. Calif. Fish, Game 62(3):179-186.

_____. 1976b. Inland fishes of California. Univ. Calif. Press. 405 pp.

_____. 1976c. Fish introductioons in California: history and impact on native fishes. Biol. Cons. 9:101-118.

Moyle, P. B. and R. Daniels. Ms. Distribution and ecology of fishes of the Pit River system in California.

Murphy, G. I. 1948. Distribution and variation of the roach (Hesperoleucus) in the coastal region of California. M.A. thesis, Univ. Calif., Berkeley.

Peters, J. C. and W. Alvord. 1964. Man-made channel alterations in thirteen Montana streams and rivers. Trans. 29th North Amer. Wildl. and Nat. Resour. Conf. p. 93-102.

Platts, W. S. and W. F. Megahan. 1975. Time trends in riverbed sediment composition in salmon and steelhead spawning areas: South Fork Salmon River, Idaho. Trans. 40th North Amer. Wildl. and Nat. Resour. Conf. p. 229-239.

Vanicek, C. D. 1970. Distribution of Green River fishes in Utah and Colorado following closure of Flaming Gorge Dam. S. W. Nat. 19(3):297-315.

White, R. J. and O. M. Brynildson. 1967. Guidelines for management of trout stream habitat in Wisconsin. Tech. Bull. No. 39, Wisc. Dept. Nat. Resour., Madison, Wisc. 64 pp.

Whitney, A. N. and J. E. Bailey. 1959. Detrimental effects of highway construction on a Montana trout stream. Trans. Amer. Fish. Soc. 88(1):72-73.

Chapter 9

CREATING LEGISLATION TO STUDY AND PROTECT CALIFORNIA'S RIPARIAN FORESTS

Senator John Dunlap
Fourth District California State Senate

I wish that I could spend 15 or 20 minutes talking solely about state legislative proposals dealing with the protection of Riparian Habitat, but there just aren't enough of them to describe. That fact itself says a lot. I am going to speak today about the "practical politics" of preserving worthwhile elements of the environment and about some of the legislative history that has taken place during the last 10 or 11 years to illustrate what is going on now.

Until the late 1960's there wasn't a great deal of legislative action to preserve our natural heritage in any particular way. In recent years, however, there have been several major accomplishments. The San Francisco Bay Conservation and Development Commission is perhaps one of the first and most important agencies formed to preserve the San Francisco Bay. It was established by temporary legislation which became permanent legislation in 1969 and 1970 after a 4 year trial period. The Tahoe Regional Planning Agency was the result of similar legislation to preserve Lake Tahoe. Legislation creating the Coastal Conservation Commission illustrates how long some conservation processes can take to become law! Special legislation which has evolved during the last few years will actually put money into preserving and supporting a moratorium on certain kinds of development in the Suisun marsh. In 1975, Assemblyman Charles Warren introduced Assembly Bill 15 to preserve agricultural land throughout the state of California. That battle is still going on; AB 15 lost in the Senate Finance Committee on the closing day of the 1976 session. The question of agricultural land preservation is back again now with more than five bills on the subject.

The interesting thing about all these particular pieces of legislation is that they are just parts of the whole project. We haven't done the job as logically as we should; we've done it on a piecemeal basis, here and there. We consult with those of us who may be interested in particular features of our natural heritage which we think will capture the public imagination and the public will, and move in that direction. That's why we work on saving San Francisco Bay, Lake Tahoe, Suisun Marsh, and some portions of the California coastline. Because these were specific regions, it was not too hard to capture the public interest and imagination over a period of a few years.

Back in 1969, Assemblyman Edwin Z'berg introduced a bill to create a general environmental agency to protect all environmental features of California. It would have been logical; it would have been efficient; it would have been binding to all existing organizations on all environmental issues. Z'berg's bill, as logical as it was and as efficient as it would have been did not even get out of his own committee. It didn't get to first base. So I think that shows you have to look around and find something that can easily be demonstrated to have public support. This may not be true forever; this may be changing as time goes by and we may be able to learn one of these days to be involved in a more logical approach, but I don't think we've quite reached that time. I think we probably still have to proceed on the least logical basis.

Bear that in mind when thinking about riparian habitat. I believe it is important to preserve riparian habitat for people who are interested and want to enjoy it; but that's not the way to do things if you were to approach a legislative committee. Legislators would have to respond to all sorts of opposing issues, and you wouldn't get very far if you were to give a "poetic" argument like that. Therefore, I think it is vitally important to those involved and interested in preserving riparian habitat to find help from those who want to preserve it for practical purposes. For example, the Suisun Marsh is an area where there is heavy wildlife production and a vital source of food supply for many animals, including man. This is a good practical reason to protect the marsh. I think we also need to include recreation value as a practical use of land. Some of us overlook the fact that though many farmers naturally want to be able to sell their land when the time comes for non-farm purposes and cash in for the large dollar, there are others who have real interest and concern for our heritage. I think, in terms of taking care of the land, we shouldn't overlook the farming community as a source of help. Another source of support could be the newly formed California Tax Reform Association. This association is not a land owners' tax reform organization, but one which is oriented to low income groups as opposed to big taxpayers. I think somewhere along the line some point might be made to this association that a great many of the Army Corps of Engineers' and other agencies' projects cost money and that they are an indirect subsidy to various special interests. The suggestion could be made that a tax reform organization should not just be interested in tax reform but also in economy and government. Therefore, it would also be interested in avoiding major projects which involve what I might call "welfare invested" interest. Editor's note: Shasta County has since adopted a riparian vegetation ordinance.

There are various existing state and federal agencies which have an interest in riparian habitat and which could be of help. Obviously, the Department of Fish and Game, the Federal Bureau of Outdoor Recreation, and other agencies such as county governments can be helpful if you find the right people at the right time and the right place. Napa County, one of the counties I represent, is the only county in the state that has a riparian stream ordinance. This is something that other counties might establish.

To illustrate how really tough it is to get strong environmental legislation through, I would like to briefly trace the course of coastal legislation in California. The problems of protecting California coast first came to public attention in 1968 when a sub-division known as the Sea Ranch was built in Sonoma County. It is a nice place to go, if you happen to have a house up there. However, the developers were in the process of sub-dividing about 14 miles of California coastline without establishing access in any one place so that the people of California could use the beaches they own! Sonoma county permitted this sub-division without requiring public access. Of course, it seemed not to be a very good idea to me, so, in the middle of 1968, I introduced a very simple bill that said when a city or county permits a sub-division on coastal land, the developer should be required to make reasonable access for the public to the California coastline. I got that bill through the Assembly in 1968 to have it defeated in the Senate Business and Profession Committee by one vote. It went through the Assembly again in 1969, then lost in the Senate Governmental Efficiency Committee. I don't know how many votes I did or didn't have because that was in a time when they took secret votes the night before at a restaurant, and fortunately those days are gone and have been gone for several years! Finally in 1970, I ran it through the assembly and got the bill channeled into the Local Government Committee instead of one of the other "killer committees"; it got through that committee and off to the floor of the Senate. It was defeated 18 to 12. It was about 3 weeks before the closure of the session and I had been at it for 3 years. I began a letter writing campaign all over the State of California. Two days before the legislature adjourned, the bill was reconsidered. As a result of the citizen action of 2,000 people and a couple of newspaper editorials in particular districts, the bill won with 22 votes and became law in 1970. Though it became law 2 years after I introduced it, it should have been law 2 years before I introduced it.

That's just Chapter 1; Chapter 2 began in September 1970 when Senator Alan Cranston and I indicated that we were going to the Planning and Conservation League Congress in Santa Barbara and that we intended to introduce plans for a full Coastal Conservation Commission. We began that bill in 1971; it got through the Assembly and was killed in the Senate. So we tried again in 1972; it passed through the Assembly and then the Coastal Alliance developed an initiative petition which was placed on the state ballot. Finally, in November of 1972, the people of California adopted what then was Proposition 20 of the interim coastal legislation. Between the time Senator Cranston and I made our announcement in the Fall of 1970 and January of 1973 when the moratorium took effect, 2 years and 4 months had passed. I point that out just to indicate how really tough it is to pass some environmental legislation and how there's a time to start things and a time not to start things. You should think in terms of being realistic and being ready to go. Draft legislation to study riparian habitat has been written and will be introduced into the Senate early in 1978. It is Senate Bill 932. Basically what it does is declare that the riparian habitat is vitally important for various reasons, and that it is a disappearing resource. It authorizes the Department of Fish and Game to spend the sum of $150,000 for the purpose of coordinating a state-wide survey of existing riparian habitats. There are studies on parts of the upper Sacramento River but there are many more areas in this state we don't know about that aren't so well charted.

As I close, I encourage you to participate in the political process. Next year, there are going to be 20 senators elected; there are going to be 80 assemblymen elected; the governor is going to be campaigning for re-election; congressmen will be campaigning for re-election and I'll be running. I suggest those who are interested in the political aspects of riparian habitat might also be interested in those elections. If you find someone in office who you don't think is responsible to this particular need, try and find somebody else. Changes are necessary. Sometimes, of course, those already in office can be educated.   There are always several things that are particularly important to listen to, but if somebody working for us and helping us becomes involved in something, we show them our interest in that area.

Here are just a couple of political "hints." Letter writing, talking to state legislators, and being with us to help with the timing for legislation is important. It's perhaps a little helpful to get letters from all over the state on a particular issue, it certainly calls attention to it, but not as much attention as a letter from somebody in the district a particular legislator happens to represent. He knows those are the people who vote for him, those are the people who work for him, and he is going to be far more responsive. I think it's also important to get to the legislators _early_ before, for some reason, we hear from "the other side." Get to them early and, before the bill comes up before their committee or for a vote on the floor, get to them again. Don't let them forget. Once in a while we get 100 postcards which are all identical and they say vote "no" on such a bill or "yes" on such a bill and they are signed. Obviously somebody took these postcards to a meeting, passed them out, and everybody signed them. Well, we're not nearly as impressed by 50 letters like that as by 5 letters that are individually written and in a person's own words. We know it is easy to take postcards to a meeting, but it doesn't necessarily mean the people who signed them are seriously interested. They probably won't watch the bill to see what happens next. So bear that in mind. I think letter writing campaigns on particular issues can be far more effective. It doesn't hurt to add a little sentence at the end of the letter stating "I would appreciate your taking the time to respond to this matter." You didn't come here to learn to be politicians, but I'm asking that you join with me.

Senate Bill No. 932
Introduced by Senator Dunlap
April 12, 1977

Legislative Counsel's Digest

SB 932, as amended, Dunlap. Riparian habitat: survey and report.

There is no existing law which provides generally for a survey and analysis of the status of the riparian plant and related wildlife community in the state.

This bill would require the Department of Fish and Game to determine the current status of the riparian plant and related wildlife community in the state and to report thereon to the Legislature by January 1, 1979. The bill would require the report to include specified matters and would make legislative findings and declarations.

The bill would appropriate $150,000 to the department for performance of such survey and analysis and the preparation of such report.

The people of the State of California do enact as follows:

SECTION 1. The Legislature hereby finds and declares as follows:

(a) The riparian plant and related wildlife community is a natural resource of vital importance to people of California.

(b) Riparian vegetation which appears as a green belt along permanent or intermittent watercourses is one of the most valuable habitats known to man, hosting an outstanding assemblage of wildlife species. Of the many habitat types occurring in California, riparian habitat provides living conditions for a greater variety of wildlife than any other type.

(c) Riparian vegetation is critical to the survival of several species of related wildlife classified as rare or endangered under authority of the California Endangered Species Act of 1970.

(d) The estimated 347,000 acres documented in the 1965 Fish and Wildlife Plan has since been further reduced by man's alterations and development of land and such reduction is continuing.

(d) In order to properly protect and manage this vital natural resource and to determine what actions are necessary for effecting such protection, a comprehensive survey and analysis of riparian habitat is necessary.

SEC. 2. The Department of Fish and Game shall determine the current status of the riparian plant and related wildlife community in the state and shall report thereon to the Legislature on or before January 1, 1979. Such report shall include an estimate of the historical riparian habitat acreage, current acreage, a description and assessment of the various types of riparian plant and related wildlife communities, an identification of existing and potential threats to the resource and a description of current problems of resource maintenance and protection. The report shall also recommend to the Legislature what actions should be taken to protect, maintain, and, where possible, enhance riparian habitat.

SEC. 3. The sum of one hundred fifty thousand dollars ($150,000) is here by appropriated from the General Fund to the Department of Fish and Game for the performance of the survey and analysis and the preparation of the report required by this act.

Chapter 10
ENVIRONMENTAL APPLICATIONS IN CORPS OF ENGINEERS WORK
WITH REFERENCE TO RIPARIAN VEGETATION MANAGEMENT

Fred Kindel
Chief of the Environmental Planning Section
Sacramento District
United States Army Corps of Engineers

Two aspects of the Corps activities in which environmental applications are important are: protecting the Sacramento Valley levee system with rock bank protection and projects for which plans have been developed to protect riparian trees and vegetation.

A system of about 1,000 miles of levees has been constructed to provide flood protection to about one million acres and about 800,000 persons living in the flood plain of the Sacramento River (Environmental Statement, 1972). The levee system is threatened by continuing erosion, and normal maintenance and even emergency measures are not adequate to cope with the danger to the levees (Sacramento River Flood Control Project, 1960). In 1960 at the request of the State of California, Congress authorized the Sacramento River Bank Protection project to protect the levees.

Although there are some variations in the work, the usual circumstance is that erosion has progressed into or near the levee which is in danger of failure. To provide protection, a section of levee is prepared by sloping to a 1 on 2 or a 1 on 3 slope and placing the rock bank protection. All trees and vegetation in the area to be rocked must be removed to slope the bank to retain the rock. In the past, trees and vegetation were also removed from some areas adjacent to the actual worksite to facilitate equipment operation while the rock is being placed.

The following design changes were made in recent years to reduce the environmental impact of bank protection work (Environmental Statement, 1972; Bank Protection General Design, 1974):

Where feasible, contractors have been required to avoid disturbing any significant vegetation outside the limits of where the rock is placed. Besides careful equipment operation from the top of the levee, work is sometimes accomplished from a barge on the river which avoids unnecessary disturbance of vegetation to the maximum. However, barges can only navigate the deeper reaches of the river south of Colusa.

Trees have been surveyed and evaluated at the edge of the bank protection areas and all individual trees which would not interfere with construction and could be saved are marked.

At some erosion sites there is still some berm area remaining between the river and the levee. By placing rock only to the top of the berm, three things are accomplished: erosion is arrested and the levee is protected; the berm is protected, permitting vegetation growth; and there is much less rock required for construction. Protecting the berm means that trees and other vegetation on the berm will not have to be removed. Placing rock only to the top of the berm means there is much less visible rock when the river is at low flow. This appears to be the most desirable of the protection methods for environmental application. More of this type of work could be done if additional funds were available (with only limited funds, work is restricted to the critical erosion sites and the other protection methods are utilized).

At some locations, the circumstances of the erosion and other factors led to a different design than adding rock for protection of the levee. Where more economical, the existing levees may be set back or relocated further from the river bank. Rock protection is placed on the riverside of the new berm, and vegetation may be planted on the berm. An example of this type of design is at a location near Monument Bend located on the right bank about one mile upstream from the Interstate 880 bridge crossing.

As each unit of bank protection is completed and turned over to the State for operation and maintenance, a supplement is provided to the standard operation and maintenance manual which covers specifically the operation and maintenance needs of that unit. Where measures are instituted for added vegetation in our construction work, it is required that this vegetation must be properly maintained.

On berm areas where there are significant trees and vegetation, the Corps has stipulated that the protected trees should remain when such sites are provided with bank protection. The State Reclamation Board has adopted a program of acquiring a stronger easement than solely for flood control purposes; this provides the landowner a higher price and requires him to leave the native riparian vegetation in place. This is an important companion feature to the berm protection design change (Bank Protection General Design, 1974).

Over the past several years, a number of experimental measures have been tested. The experimental program has had two primary purposes: to test the effectiveness of alternative bank protection methods and materials, and to determine costs of such alternative methods. The testing has been to determine engineering and economic characteristics on the effectiveness of the alternative methods as well as their environmental contribution. One important factor is whether alternative or supplemental methods are more costly to operate and maintain. Where possible, alternatives should be found that do not add significant maintenance expenses.

A pilot levee maintenance study was conducted by the State of California and reported on in 1967 (Pilot Levee Maintenance Study, 1967). The study demonstrated that certain types of ground cover were compatible on levees, that some trees and shrubs may be allowed on some levees, and that in most cases unrestricted growth may be allowed on berms. The study indicated that costs of maintenance of levees would be increased with this vegetation.

The Corps has planted trees and shrubs at several selected sites along the Sacramento River (Environmental Statement, 1972) to demonstrate that such vegetation can be successfully grown, can be compatible with flood control requirements, and can offer a significant improvement to aesthetics and other environmental aspects of the river. The most outstanding example of such a demonstration is near Monument Bend just upstream from Interstate 880 bridge. In 1967 we planted a variety of trees and shrubs along about three miles of the riverbank where the levee had been set back and the new berm protected by rock. In 1970 after three years, the vegetation had provided a significant improvement (Environmental Statement, 1972) and this is still in evidence today. The State Department of Water Resources conducted some maintenance studies on this vegetation demonstration site and in 1973 reported on the survival rates of the various species in relation to the effects of inundation by floodwaters and accidental losses by fire. Cost of manpower for levee maintenance with the planted vegetation was increased by 64 percent over costs without vegetation on similar adjacent levee areas (Sacramento River Levee Revegetation Study, 1973).

The Sacramento River and Tributaries Bank Protection and Erosion Control Investigation, authorized by the House Public Works Committee, was initiated in 1977. The purpose of this study is: To determine the Federal interest in, and responsibility for, providing bank protection and erosion control; to study alternative means and the feasibility of providing a comprehensive program to stabilize the streams, protect the levees and banks, preserve riparian vegetation, wildlife habitat and aesthetic values, and provide outdoor recreation opportunities along the river; and to select and recommend the best and most balanced plan of improvement, provided that such a plan is found feasible. Completion of the study is scheduled for 1982.

Literature Cited

Bank Protection General Design, Design Memorandum No. 2, Sacramento River Bank Protection Project, December 1974.

Environmental Statement, Sacramento River Bank Protection Project, Sacramento District, Corps of Engineers, November 1972.

Pilot Levee Maintenance Study, Bulletin No. 167, Department of Water Resources, State of California, June 1967.

Sacramento River Flood Control Project, California, Senate Document No. 103, 86th Congress, 2nd Session, 26 May 1960.

Sacramento River Levee Revegetation Study, Department of Water Resources, State of California, July 1973 (Central District, D.W.R.).

Chapter 11

THE UPPER SACRAMENTO RIVER TASK FORCE: A PROGRESS REPORT

James W. Burns
Assistant to the Resources Secretary of California

About a year and a half ago, the Secretary for Resources established the Upper Sacramento River Task Force. It was made up of the following departments: Navigation, Fish and Game, Water Resources, and Parks and Recreation. Also represented were the Wildlife Conservation Board, the State Reclamation Board, the Water Quality Control Board and the State Lands Commission. When we first met, it was realized that there was a lot of construction along the river over which we didn't have full jurisdiction. Therefore, we invited federal and local governments to join us. Now, our membership includes the Bureau of Reclamation, Army Corps of Engineers, Bureau of Land Management, Fish and Wildlife Service and the Bureau of Outdoor Recreation. Shasta, Tehama, Glenn, Colusa, and Butte counties are also represented.

The objectives of the task force were to coordinate the activities of the many governmental agencies which have jurisdiction over developments along the Sacramento River and to insure protection of the fish and wildlife and recreational aspects of the river. We also had to consider the use of river water for irrigation and the use of levee land for agriculture. The Task Force's first job was to identify the areas along the river which had resource conflicts. Many problems dealt with fisheries and other water related activities, but what I'll deal with today is the problem of riparian vegetation removal.

After identifying the problems, we developed a list of actions we should take that would help to protect the riparian vegetation. Our first effort was to make local governments aware of the problem. Then, we looked at Senator Dunlap's area in Napa county where an ordinance allows for the protection of riparian vegetation along the water courses in that county. We changed it a bit, made a permit system and added a general plan element which would control removal of riparian vegetation. Most counties looked at the ordinance and rejected it. A few counties began looking at riparian vegetation protection as part of their flood plain zone, but several local government officials said riparian vegetation protection wasn't in accord with the flood water capacity and it would impair flow through the canals. Counties with set back levees have riparian vegetation between the levees and the rivers, and we plan to have another ordinance drafted which will list what land uses are compatible and incompatible within the levee systems.

We haven't given up on the counties yet. "Local rule" is an issue here, and we want to give local people every chance to adopt plans for preservation, protection, and re-establishment of riparian vegetation. After we've exhausted that approach, we'll take our own legislative actions.

With regard to the Army Corps of Engineers, we encouraged them to evaluate their bank protection and levee maintenance projects and to ask Congress to authorize mitigation for the Sacramento River bank protection project. The Task Force suggested that they include the following in their study: (1) an objective appraisal of results of past bank protection efforts, (2) the investigation of alternative bank protection methods, (3) a study of bank protection techniques and construction methods which minimize adverse impact on water quality, fish and wildlife, and scenic values, and (4) a list of specific criteria for determining where bank protection should be applied and how priority should be established.

Recently, we've been negotiating on the Chico Landing to Red Bluff Project, and we've been able to get the land owners to agree to some environmental considerations. The State Reclamation Board established that in areas already under agriculture there will be a 30 ft. corridor where riparian vegetation could be re-established. In areas that are presently riparian vegetation, the corridor is 150 ft. The Board will have more direct control over treatment of future bank protection sites, and it is adopting a policy that "recognizes the vital importance of riparian vegetation for fish and wildlife, recreation and study quality." The Board's intent is that all activities under its permits include recognition of the value of riparian vegetation consistent with flood control purposes and that actions be taken to conserve and encourage riparian growth. The Secretary of Resources has asked the

Army Corps of Engineers to prepare separate environmental impact statements for any future bank protection work. These statements are to be "site specific." That means a new report for each project, not just a rewrite of previous statements. The Corps is also to seek authorization to include mitigations in all future bank protection work, the cost of which will be included in the project.

Our next effort is one that's going on right now, and that's to get the State Board of Forestry to regulate timber harvest in riparian lands in the Central Valley. To do this the Board must designate all riparian lands as timber lands and riparian tree species as commercial species. This proposal to the Board of Forestry is presently under review by the Resources Agency, and we hope to send it to the Board within the very near future.

I certainly hope that a joint effort between local and state government agencies and public groups such as Audubon Society, Sierra Club and the Sacramento County Landowners' Association will insure riparian vegetation protection along the Sacramento River and will set an example of what can be done on other rivers throughout the state.

Chapter 12

A NONDESTRUCTIVE APPROACH TO REDUCING RIPARIAN TRANSPIRATION

David C. Davenport
Department of Land, Air and Water Resources:
Water Science and Engineering Section
University of California, Davis, California 95616

Most of California's water is gathered from precipitation on the watersheds. This water, via surface and subsurface runoff, normally passes through riparian zones en route to utilization by agricultural, industrial, and domestic users, eventually reaching the ocean. The guaranteed supply of surface and/or subsurface water in riparian zones has led to the establishment and proliferation of vegetation, the species varying chiefly with climate (determined mainly by altitude and latitude), rooting characteristics, and other adaptive features. The accessibility of water to the root zone usually results in vegetative canopies which transpire in general proportion to their spatial density. The vegetative density and plant species are determinants of the population and types of wildlife which inhabit or visit the riparian zones.

Plants growing along streams and rivers which flow most of the year are termed riparian vegetation. Meinzer (1923) initiated the term "phreatophyte" from two Greek words meaning "well plant", i.e., plants whose roots tap the groundwater table or the wet soil just above it. Most riparian vegetation is therefore phreatophytic. The phreatophyte zone, however, also includes the deeper alluvium beyond the riparian zone where roots penetrate to the groundwater table for additional water (Horton, 1974). Over 70 plants in the western United States are classified as phreatophytes (Robinson, 1958). These include alder (Alnus sp.), cottonwood (Populus sp.), greasewood (Sarcobatus sp.), mesquite (Prosopis sp.), oak (Quercus sp.), quaking aspen (Populus sp.), Russian olive (Elaeagnus sp.), saltcedar (Tamarix sp.), saltgrass (Distichlis sp.), seepwillow (Baccharis sp.), sycamore (Platanus sp.), and willow (Salix sp.). Todd (1970) listed phreatophytes and their areas and consumptive use in river basins of the western U. S.

The most important water-consuming riparian phreatophytes in the west are cottonwood, saltcedar, and willow. Over 16 million acres (about 6.5 million ha) of phreatophytes in the western states annually transpire over 25 million acre-feet (about 31,000 million m^3) of water (Robinson, 1958). Robinson (1965) also estimated that the aggressive saltcedar occupies over 1.3 million acres (526,000 ha), transpiring over 5 million acre-ft (6 million m^3). Water use by saltcedar varies from 2 to 8 ft. per year, depending on location, density, salinity, and depth to the water table.

The earlier literature reveals the controversies that have developed on the removal of forest vegetation, the hydrologic concepts involved, and the environmental impacts expectable. Until the early part of this century it was believed that forests had no material influence on stream flow (Chittenden, 1909). About two decades later, Hoyt and Troxell (1934) strongly disputed the idea that forests "conserve water". They cited data from: 1) investigations during the period 1910-1926 by the U. S. Forest Service and Weather Bureau (Bates and Henry, 1928) at Wagonwheel Gap, Colorado, in which vegetation removal increased annual streamflow by 15% and maximum daily flow by 46%; and 2) their own studies in southern California, where a chance wildfire in 1924 that destroyed the vegetation, increased annual streamflow by 29% and maximum daily streamflow by 1700%. Although humus from forest vegetation was acknowledged to improve the water infiltration and storage capacity of soils and to reduce erosion, Hoyt and Troxell firmly stated that the widely held belief that forests increase summer runoff was a fallacy. Deforestation in the Wagonwheel Gap area did not increase erosion because direct surface runoff is normally small there, but in the burned-over southern California watershed erosion was clearly evident. Those workers stated that they are "lovers of forests and have a keen appreciation of the value of water. . . and of forests as a source of wood. . . and as playgrounds. . ." Because forests are users, not conservers, of water, they concluded that "in regions where (water) supply is a controlling economic factor, careful study is needed to determine whether the value of increased water supply and better sustained minimum flow which are shown to obtain without forests, does not outweigh the benefits of lowered normal flood flows and decreased erosion produced by forests, especially if these benefits can be obtained by shrubs or other small growth without the loss of water occasioned by forest growth."

Hoyt and Troxell's paper generated a flurry of responses from eminent foresters, engineers, and hydrologists, mostly supportive of the idea that forests are users, not conservers, of water. Nevertheless, one of the foresters, C. G. Bates, indignantly wrote that the tone of the Hoyt and Troxell paper was "directed against foresters who see in the forest cover a beneficial effect." Implying (erroneously) that Hoyt and Troxell were recommending indiscriminate destruction of vegetation, Bates declared: "It seems scarcely probable that even the people of Southern California would place such stress on increased population and increased water supplies that they wouldchoose to have their entire foothills region assume the barrenness of the upper slopes of Mt. Wilson. There may be communities in dire need of additional water, but it is not believed there are any that can afford to obtain it by a method so wholly destructive in its design. Has the time arrived when man may assume the audacity to start reversing these great constructive processes of nature, for the sake of a temporary benefit, to water a few paltry additional acres of crops or to permit a few more people to congregate in the cities?"

The Wagonwheel Gap and the Southern California experiments spurred investigations on the effects of forest removal on streamflow in the eastern U. S., where rainfall is more abundant. Hoover (1944) estimated a water use of 17-22 inches per year by forest cover with a dense shrub understory on the Coweeta Experimental Forest in the southern Appalachain mountains. Removing the vegetation from a 33-acre watershed, while leaving the humus cover intact to maintain infiltration and reduce erosion, increased annual streamflow 100% if vegetative resprouting was prevented. On the other hand, Dunford and Fletcher (1947) demonstrated that cutting only riparian vegetation along the stream bank of a 22-acre Coweeta watershed immediately eliminated diurnal fluctuation in water flow (which occurs because of daytime transpiration by riparian vegetation). They suggested that the daily flow increases of 4-19% due to the cutting can be of practical value on municipal and industrial watersheds during drought years.

On the other hand, removal of vegetation to increase water yield is not always beneficial. Wilde et al. (1953) found that clear-cutting a stand of aspen in Wisconsin raised the water table 14 inches, converting a fairly well-drained soil into a semi-swamp and increasing runoff, erosion, and road damage during the rainy season. The type of impact from vegetation removal thus depends on site conditions and season. Rowe (1963) emphasized that the substantial increases in water yield observed after clearing woodland riparian-vegetation from selected areas bordering stream channels in southern California were particularly important because they occurred mainly in summer and in the initial period of soil wetting in subsequent rainy seasons, when streamflow was lowest and water most needed. When rain was heavy and the soil was wet, however, vegetation removal had no effect on streamflow. The only adverse effect on water quality noted by Rowe was an increased stream algae content after vegetation that shaded the streamcourse was removed.

Young and Blaney (1942) summarized several earlier studies on water use by native vegetation, including "canyon-bottom" (riparian) vegetation in southern California: streamflow measurements at two sites in Temescal creek indicated water use by the moistland vegetation was 0.24 - 0.66 inches/day in May, 1929; in the Santa Ana Valley, willows were estimated to transpire 45 inches/yr; in Coldwater Canyon mean daily water use by vegetation (mostly alders) was 0.35 acre-inches (nearly 10,000 gallons) per 1000 feet of canyon bottom.

This review of studies on water use by riparian vegetation indicates that considerable quantities of water (roughly 4-6 feet/yr.) can be lost to the atmosphere, thereby depleting streamflow. Because of increasing competitive demands for water, particularly in summer, when both demands and transpiration losses are highest, people are faced with priority decisions between their need for the water transpired by riparian plants and the ecological implications of eradicating the vegetation (e.g., effects on wildlife, erosion, water quality, etc.) Compromise solutions could involve partial removal of riparian vegetation or a method of curtailing transpiration without destroying the vegetation.

There is a need for accurate information on: 1) the type, density, extent, and water use of riparian vegetation; 2) groundwater and its seasonal variability; and 3) land ownership of riparian zones. Affleck (1975) suggested that systematic surveys using aerial and sattelite photographs supplemented by ground data would permit a more knowledgeable and equitable allocation of riparian vegetation zones for water yield improvement while protecting wildlife, recreation, and aesthetic purposes. He also suggested that practices to improve water yields should be used only where the predominating riparian species are the heavier users of water.

Approaches to reducing transpiration from riparian zones and environmental effects

Information on water yield improvement by vegetation management has recently been reviewed by Ffolliott and Thorud (1975), Affleck (1975), and Horton (1976). The methods reviewed include: 1) eradication by burning, mechanical and/or chemical treatments; 2) thinning of cottonwood stands; 3) conversion to other vegetation; 4) channelization to reduce the area of flood plain having a high water table; 5) biological control by introduction of insect and plant pathogens to limit the growth and spread of saltcedar; and 6) antitranspirant (AT) treatment. The adverse environmental impacts of these methods range from severe (eradication) to moderate (conversion or thinning) to minimal (AT).

Eradication can lead to loss of wildlife habitat, herbicide contamination, lowered aesthetic quality, and increased erosion. Complete clearing of riparian vegetation will not necessarily conserve all the water lost to transpiration. Increased evaporative losses from bare soil and open water and transpiration losses from replacement or re-sprouted vegetation reduce the water salvage potential of clearcutting. Thus, after the U. S. Geological Survey cleared phreatophytes (mostly saltcedar and mesquite) along the Gila River, Arizona, there was a saving of only 30 inches of the original (preclearing) 50 inches of consumptive use (Hanson et al., 1972). Channelization destroys the natural character of a river; it may be more justified for flood protection than for water salvage.

Thinning cottonwoods in Verde Valley, Arizona, was discontinued in 1969 partly because of adverse public opinion on the effects on wildlife and aesthetics, and insufficient evidence of improved water yield. Johnson's (1970) studies in Verde Valley showed that the numbers and diversity of bird species were greatest in areas of densest riparian vegetation. Thinning cottonwood trees from 47 trees/acre to 26 (moderate) and 10 (severe) trees/acre, respectively, reduced the pairs of nesting birds by up to 27% and 56%. Van Hylckama (1970) demonstrated that thinned-out stands of saltcedar use nearly as much water as unthinned stands. Trimming the plants to 18 inches high reduced evapotranspiration 35% initially, but the savings soon diminished with rapid regrowth of the saltcedar shoots (at 5 cm/day).

Campbell (1970) suggested that managing riparian vegetation in the southwest to increase water yield may require selective clearcutting, rather than complete eradication "to maintain a biological balance and thus prevent thermal pollution, channel erosion, and destruction of aquatic and wildlife habitats." Phreatophyte areas provide shelter and nesting grounds for white-winged and mourning doves in the Arizona flood plains (Cottam and Trefethen, 1968), although the main source of food is often from nearby agricultural fields. Before saltcedar invaded the lower riparian zones of the southwest in the 1930s, the mesquite bosques provided both shelter and food for the doves (Arnold, 1943). Clearing the mesquite depleted the dove population. Horton and Campbell (1974) suggest that a compromise be made between preserving cover for nesting of doves and removal of cover to produce food for waterfowl.

Beekeepers have been deeply concerned about projects to clear riparian vegetation which provide bee pasture. This is particularly important in areas where, and at times of year when, the phreatophyte bloom is the only or the major source of nectar and pollen to bees. An advantage of the AT approach to reducing riparian transpiration is that the sprays can be timed to avoid the flowering period but coincide with the period of maximum vegetative cover, when transpiration rates are highest.

Antitranspirants

There are two major groups of AT: stomatal inhibitors (Zelitch, 1965); and "film-forming" materials (Gale and Hagan, 1966). Brooks and Thorud (1971) and Cunningham and Thorud (1971) examined the effects of the former type on saltcedar. Davenport et al. used the latter type on various phreatophyte species, sprayed from the ground (1976a) and by air (1976b). Those workers all showed the potential of ATs for water conservation from phreatophyte vegetation. The factors determining the effectiveness of ATs and their application in hydrology were reported by Davenport et al. (1969). Although currently available ATs curtail both photosynthesis and transpiration, a temporary decrease of photosynthesis in nonagricultural phreatophyte vegetation is probably of little importance compared with the saving of water without destruction of the plants. The ingredients of some "film-type" ATs already have EPA clearance and have therefore been used to increase yields of certain food crops (Davenport et al., 1974). However, the biological effects of all antitranspirant materials on insects, animals, and aquatic life have not been extensively investigated. Studies of the effects of a wax-based AT on some index species of aquatic and terrestrial wildlife in phreatophyte areas are currently being made under OWRT Project C-6030 CAL involving the Universities of California, Colorado, and Arizona (1976).

The antitranspirant (AT)aapproach therefore seeks to curtail transpiration from forested watersheds (Belt et al., 1975; Davenport et al., 1969; Hart et al., 1969; Waggoner and Turner, 1971) and riparian vegetation, such as saltcedar (Brooks and Thorud, 1971; Cunningham and Thorud, 1971; Davenport et al., 1976a) without removing the vegetation or damaging the ecological balance. This approach was suggested also by the National Academy of Sciences Committee on Technologies and Water in their Report to the National Water Commission (1971). Recent reviews by Horton (1976), Ffolliott and Thorud (1975), and Affleck (1975) have indicated the potential usefulness of spraying suitable ATs as an ecologically acceptable means of curtailing water use by phreatophytes. Although currently available ATs are expensive (partly because of an undeveloped market for this relatively new technique and inadequate research to improve their efficiency), Affleck points out that the ". . .cost may be justified because wildlife, recreation, and aesthetic values of the riparian zone would be left undisturbed." He further states that ". . .antitranspirants provide greater flexibility in their use compared to other methods. They could be applied at times when transpiration rates are highest, in areas where water is needed most, or not applied at all during wet years when there is no need for increased water yields. Another benefit of antitranspirants in comparison to eradication methods is that the riparian canopy remains intact, thus shading the soil surface and reducing possible soil evaporation losses. If antitranspirants are proven to be effective and safe, the development of this management alternative may be mutually acceptable to water, recreation, and wildlife interests."

The Department of Water Resources, California (1976), emphasizes efficient use of existing water supplies, particularly during the current drought. Reducing riparian transpirational losses of limited streamflows during the first half of the summer of this 1977 drought year may not only improve streamflow, but in some cases, also extend the availability of limited water for maintaining the riparian species until the next winter rains.

Transpiration and antitranspirant studies by the University of California, Davis

At Davis California, a wax-based AT emulsion ('Mobileaf FG', Mobil Chemical Co.) sprayed on foliage of potted saltcedar, cottonwood, and willow reduced transpiration by 35-75% initially and 17-56% after four days (Davenport et al., 1976a). Gas-exchange studies on branches in a natural stand of dense saltcedar bushes along Cache Creek, California, suggested that the AT conserved water much more effectively when applied to the outer part of the canopy than to the inner shaded foliage, where transpiration was already minimal. This suggested that aerial spraying of AT on the outer foliage of riparian canopies would effectively reduce transpiration. Trials with a fixed-wing plane near Lahontan Reservoir, Nevada, and a helicopter at Cache Creek, California, provided information on AT spray distribution in canopies (Davenport et al., 1976b). Scanning electron-microscope photomicrographs showed the distribution of AT wax on the foliar surfaces (Fisher and Lyon, 1972). Turbulence from the aircraft aids spray distribution in the canopy and coverage of the stomata bearing surfaces of the leaves. Evidence of effectiveness of aerially sprayed AT was shown by increased resistance to diffusion of water vapor from the foliage (measured by a porometer) and reduced rates of transpiration (measured by gas exchange). These measurements were made only on small twigs on the outer canopy, however, and therefore did not show the magnitude of transpiration reduction for a complete bush or the full area of riparian vegetation sprayed.

A clearer understanding of the magnitude of transpiration from complete bushes of saltcedar and cottonwood was obtained by growing them in 100 large (15-gallon) weighable drums in the field. The drums of saltcedar could be grouped to form canopies that transpired 20-50% less than isolated plants. Transpiration for a unit ground area of saltcedar varied from 2.2 (sparse-) to 15.8 (dense-stand) mm/day in July at Davis. Extrapolation of experimental water-use data to field sites must therefore be made with care, especially since plant water consumption varies not only with stand density but also with location (altitude and climate), season, soil salinity, and soil water availability (associated with depth of the water table in the case of phreatophytes). Interestingly, a moderately dense stand of saltcedar drums at Davis in July, 1976, transpired at the same rate (about 6-7 mm/day) as a natural stand of 12-15 foot-tall saltcedar in June, 1976, in New Mexico. We measured the latter in the large (1000 ft^2) U. S. Bureau of Reclamation lysimeter tanks located among similar vegetation on the flood plain of the Rio Grande River.

After the foliage of adequately watered cottonwood plants in 15-gallon drums was sprayed by a back-pack mist blower with 6% solutions of Mobileaf FG or Folicote (both wax-based AT's), transpiration was immediately reduced by about 35% without phytotoxicity, with the effect diminishing gradually to about 20% after two weeks and 5% after a month (Fig. 1). Higher porometer readings of diffusive resistance on treated plants confirmed that the wax sprays had indeed provided an effective partial barrier to the diffusion of water from the cottonwood leaves.

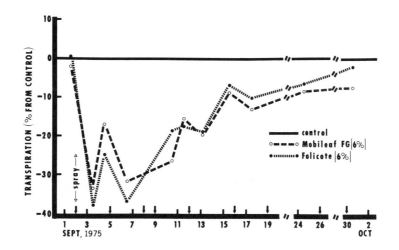

Figure 1. Transpiration reduction by the antitranspirants Mobileaf FG and Folicote sprayed at 6% (v/v) dilution with water on cottonwoods growing in 15-gallon drums. (Small arrows show irrigation dates.)

On saltcedar, which has feathery foliage and scale-like leaves, a 10% concentration of Folicote AT was required to produce a 35% reduction in transpiration, measured in the 15-gallon drums at Davis in 1975. In June, 1976, we sprayed 10% Folicote by a back-pack mist blower on a natural stand of saltcedar on the U. S. Bureau of Reclamation's 1000-ft^2 lysimeters at Bernardo, New Mexico, and again observed a 30-35% reduction in water use initially, diminishing to 10% after three weeks, with no signs of plant damage.

We conducted an experiment at Davis to determine if spraying AT from above (simulated aerial spray) on the top canopy only of a moderately dense stand of saltcedar (growing in a group of 15-gallon drums) was as effective as spraying each plant completely. Transpiration reduction by 10% Folicote was initially about 40% for the fully sprayed, and 30% for the top sprayed plants (Table 1). After a week the reduction in both cases was about 23%. However, since the top spraying required about 40% less AT than the full application, the amount of water saved per unit of AT spray applied could be 30-70% more efficient for an aerial spray covering the upper portion of a moderately dense saltcedar canopy than for a spray operation aimed at coverage of the entire canopy.

Table 1. Water savings by 10% Folicote antitranspirant ('AT') on saltcedar when spraying the plant completely (Full) at 0.7 liters/plant, or from above on only the upper canopy (Top) at 0.4 liters/plant.

Water Saved by 'AT'	8/31-9/2		9/2-9/3		9/8-9/9	
	Full	Top	Full	Top	Full	Top
% transpiration reduction	41	33	39	30	23	22
kg water saved/plant	2.03	1.64	0.89	0.67	0.41	0.40
kg water saved/liter 'AT' sprayed/plant	2.90	4.10	1.27	1.68	0.59	1.00

We also noted that saltcedar transpires at night, although at only 10% of the daytime rate. Reduction of transpiration at night by the AT is therefore of little consequence. AT effectiveness increased with: 1) a higher ratio of day-to-night hours; and 2) when soil water was not limiting. Therefore, an AT can be expected to be most effective when it is most needed for reducing water loss, i.e., on long summer days in riparian areas where groundwater is always available for transpiration. Furthermore, since transpiration transmits pure water to the atmosphere, curtailment of this loss by an AT should increase the flow of pure water to downstream users and improve water quality. The importance of this effect will vary with local conditions and is yet to be evaluated.

Regional cooperative research on a non-destructive approach to reducing riparian transpiration

Funding of this research, which is still in progress, has come from the U. S. Department of the Interior through the Bureau of Reclamation and the Office of Water Research and Technology. It involves researchers from the Universities of Arizona, California, and Colorado, and Idaho State University, bringing together a range of expertise on plant physiology, plant-water relations, micrometeorology, hydrology, and fisheries and wildlife biology. Our objectives are to: 1) determine the magnitude of transpiration from riparian phreatophyte canopies by weighing, lysimetric, energy-budget, and diffusion-resistance techniques; 2) select and test suitable ATs to reduce phreato-phyte transpiration without plant damage; 3) note the effects of ATs on the gas exchange of the plants and interactions with environmental factors; 4) refine the methods and timing of AT sprays; and 5) assess the effects of selected ATs on index species of terres-trial and aquatic life typical of those riparian locations to be treated.

Horton (1976) outlined future research needs in the moist-site phreatophyte zones, and stated: "Much more needs to be done in order to develop positive programs for optimum management of phreatophyte and riparian sites." If funding permits, we propose making a comparative analysis of alternative riparian phreatophyte management strategies for water yield and quality improvement, including environmental impacts. We realize that the costs of ecologically acceptable management strategies such as AT sprays can be high, but so also are the demands for water and for protection of the riparian environment.

Summary

The continuous availability of ground water to riparian phreatophytic vegetation re-sults in large transpiration losses to the atmosphere. These are most significant in summer when high transpiration rates coincide with peak demands for ground- and stream-water. While complete eradication of riparian vegetation can increase stream flow, it can also cause undesirable physical (e.g., erosion), and ecological (e.g., to wildlife) dis-turbances. On the other hand, spraying a non-destructive antitranspirant (e.g., a wax-based emulsion) on riparian vegetation can reduce transpiration, by over 50% in lab. tests and 30% in field tests, with no phytotoxicity. This alternative to eradication is being tested as a means of improving ground- and surface-water supplies with minimum risk of environmental damage.

Literature Cited

Affleck, R. S. 1975. Potential for water yield improvement in Arizona through riparian vegetation management. Unpublished Ph.D. dissertation. School of Renewable Natural Resources, University of Arizona, Tucson. 238 pp.

Arnold, L. W. 1943. The western white winged dove in Arizona. Ariz. Game & Fish Comm. 103 pp.

Bates, C. G. and A. J. Henry. 1928. Forest and stream-flow experiment at Wagonwheel Gap, Colorado: Final Report on Completion of the Second Phase of the Experiment. Monthly Weather Rev. Suppl. 30.

Belt, G. H., J. G. King and H. F. Haupt. 1975. Streamflow modification by a silicone antitranspirant. Watershed Management Symp. ASCE, Logan, Utah. August. pp. 201-206.

Brooks, K. N. and D. B. Thorud. 1971. Antitranspirant effects on the transpiration and physiology of Tamarisk. Water Resc. Res. 7:499-510.

Campbell, C. J. 1970. Ecological implications of riparian vegetation management. J. Soil & Water Conserv. 25:49-52.

Chittenden, H. M. 1909. Forests and reservoirs in their relation to stream flow, with particular reference to navigable rivers. Trans. Am. Soc. Civ. Eng. 62:245-546.

Cole, J., President New Mexico Beekeepers Assoc., Private communication.

Cottam, C. and J. B. Trefethen, (Editors). 1968. Whitewings. D. van Norstrand Co. Inc., N. Y. 348 pp.

Cunningham, R. S. and D. B. Thorud. 1971. Antitranspirants: a possible alternative to the eradication of saltcedar thickets. Proc. New Mexico Water Conference, Las Cruces, New Mexico. 16:101-109.

Davenport, D. C., R. M. Hagan and P. E. Martin. 1969. Antitranspirants research and its possible application in hydrology. Water Resc. Res. 5:735-743.

Davenport, D. C., K. Uriu and R. M. Hagan. 1974. Effects of film antitranspirants on growth. J. Exp. Bot. 25:410-419.

Davenport, D. C., P. E. Martin, E. B. Roberts and R. M. Hagan. 1976a. Conserving water by antitranspirant treatment of phreatophytes. Water Resc. Res. 12:985-990.

Davenport, D. C., P. E. Martin and R. M. Hagan. 1976b. Aerial spraying of phreatophytes with antitranspirant. Water Resc. Res. 12:991-996.

Dept. of Water Resources. 1976. Water conservation in California. State of California DWR Bull. 198. 95 pp.

Dunford, E. G. and P. W. Fletcher. 1947. Effect of removal of streambank vegetation upon water yield. Trans. Amer. Geophys. Union. 28:105-110.

Ffolliott, P. F. and D. B. Thorud. 1975. Water yield improvement by vegetation management: Focus on Arizona. Report for Rocky Mountain Forest and Range Expt. Sta. (Univ. Ariz., School. Renewable Resc.), 14 chapters. 1045 pp.

Fisher, M. A. and T. L. Lyon. 1972. Antitranspirant film detection by scanning electron microscopy of cathodoluminescence. Hort. Sci. 7:245-247.

Gale, J. and R. M. Hagan. 1966. Plant antitranspirants. Ann. Rev. of Plant Physiol. 17:269-282.

Hanson, R. L., F. P. Klipple and R. C. Culler. 1972. Changing the consumptive use on the Gila River flood plain, southeastern Arizona. IN: Age of Changing Priorities for Land and Water. Irrigation and Drainage Division Specialty Conference, Spokane, Washington. Sept. 26-28, 1972. Amer. Soc. Civil Engineers.

HART, G. E., J. D. Schultz and G. B. Coltharp. 1969. Controlling transpiration in aspen with phenylmercuric acetate. Water Resc. Res. 5:407-412.

Hoover, M. D. 1944. Effect of removal of forest vegetation upon water yields. Trans. Amer. Geophys. Union, Hydrology. 6:969-975.

Horton, J. S. 1974. Management alternatives for the riparian and phreatophyte zones in Arizona. Proc. 18th Ann. Ariz. Watershed Symp., Ariz. Water Commission Rep. No. 6:4042.

Horton, J. S. 1976. Management of moist-site vegetation for water. Pacific Southwest Interagency Committee Contract Report. GPO 833-655. U. S. Bureau of Reclamation, Denver. February 1. 41 pp.

Horton, J. S. and C. J. Campbell. 1974. Management of phreatophyte and riparian vegetation for maximum multiple use values. Research Paper RM-117, USDA Forest Service, Rocky Mountain Forest and Range Expt. Sta., Fort Collins.

Hoyt, W. G. and H. C. Troxell. 1934. Forests and streamflow. Trans. Am. Soc. Civ. Eng. 99:1-30.

Johnson, R. R. 1970. Tree removal along southwestern rivers and effects on associated organisms. Am. Philos. Soc. Trbk. 1970:321-322.

Meizner, O. E. 1923. Outline of ground-water hydrology, with definitions. U. S. Geol. Surv. Water Supply Pap. 494:99-144.

National Academy of Sciences Committee on Technologies and Water. 1971. Potential technological advances and their impact on anticipated water requirements. Report to the National Water Commission, Washington, D. C. 246 pp.

Robinson, T. W. 1958. Phreatophytes. Geol. Surv. Water Supply Pap. No. 1423. 84 pp.

Robinson, T. W. 1965. Introduction, spread and areal extent of saltcedar (Tamarix) in the Western States. U. S. Geol. Surv. Prof. Pap. 491-A. 12 pp.

Rowe, P. B. 1963. Streamflow increases after removing woodland-riparian vegetation from a southern California watershed. J. Forestry. 61:365-370.

Todd, D. K. (Editor). 1970. The Water Encyclopedia. Water Info. Center, Port Wash. N.Y. 559 pp.

Universities of California, Colorado and Arizona. 1976. Factors influencing usefulness of antitranspirants applied on phreatophytes to increase water supplies and enhance water quality. Ann. Rep. (1975-76) OWRT Proj. C-6030CAL.

Van Hylckama, T. E. A. 1970. Water use by saltcedar. Water Resc. Res. 6:728-735.

Waggoner, P. E. and N. C. Turner. 1971. Transpiration and its control by stomata in a pine forest. Bull. 726. Conn. Agric. Expt. Sta., New Haven. 87 pp.

Wilde, S. A., E. C. Steinbrenner, R. S. Pierce, R. C. Dosen and T. D. Pronin. 1953. Influence of forest cover on the state of the ground water table. Proc. Soil. Sci. Soc. Amer. 17:65-67.

Young, A. A. and H. F. Blaney. 1942. Use of water by native vegetation. Calif. Div. of Water Resc. Bull. 50. 160 pp.

Zelitch, I. 1965. Environmental and biochemical control of stomatal movement in leaves. Biol. Rev. 40:463-482.

Chapter 13

THE VALUE OF RIPARIAN FORESTS IN TODAY'S SOCIETY

Daniel S. Frost
Attorney and Member of the California State Water Commission

I wish that good news could be reported about what's been accomplished over the past year. Unfortunately, I don't think we heard any good news from the previous speakers with regard to concrete accomplishments. And when I say concrete, I mean the kind that are brought to us by some of our national agencies. I think all I can do is perhaps chronical some examples and make some suggestions about what can be done in the coming year. One thing is pertinent: unless people like you become active in urging those who are in government agencies to change their practices, then it will continue to be "business as usual." What we will then be able to do is meet in Chico or Davis each year and mourn the continued destruction of what is really basically the last few remnants of a magnificent heritage. I don't think the value of riparian habitat has to be reinforced in anyone that's here today. Needless to say, it's a resource that is almost unmatched in its richness and variety in California.

The experience of going to a riparian forest is, I think, unmatched in any experience I've ever had. The variety of wildlife and birdlife and being able to see those areas along the Sacramento River that still operate largely on natural principles are truly wonderful events. But they are fast disappearing. I think that for people in the private sector to effectively fight their continued destruction, it's essential that they learn first the nature of the process of destruction. I believe a great many people in the scientific disciplines understand the natural process of a stream, or a forest, but a force that people sometimes don't understand is the nature of the destructive process. If you're going to fight the destructive process, you must understand that process.

The destructive process on those areas of riparian forest which lie above stretches of the Sacramento River that have already been converted to levees come from two sources. They are the private destruction of riparian forest by land owners and government sponsored destruction of what is humanly known as "bank protection." Destruction by private land owners began when the white man first arrived, but its acceleration began in earnest when the Shasta project was built in the late 1930's and early 40's, thereby substantially reducing the instance of floods along the Sacramento River. When that was done, land that was previously and severely flooded then became more attractive to conversion for agriculture. That land almost invariably was riparian forest. In the last twenty years, those riparian forests have been destroyed wholesale. The destructive process is still going on and will continue being aided and abetted by government agencies unless we can do two things: change the mission of the destructive agencies to one of "least action" and change the mission of the balance of our agencies to one of permanent protection. The only thing we're going to have left in a very short time is a few sad remnants of our riparian forest, generally consisting of perhaps a few strips of oaks or cottonwoods along the Sacramento River, perhaps a tree or two here or there along one of the forest bank protection projects.

It's going to take political action: it's going to take people who are familiar with the programs the Army Corps of Engineers has, and believe me it has a slug of them; it's going to take a thorough understanding by members of the public of precisely the methods used by land owners and some of the local agencies and other groups to obtain the kind of public funding they need to destroy what's left of our riparian habitat. In fact, the public is subsidizing them for the destruction of that habitat. It's going to take people who are willing to go to the Reclamation Board and try to convince those people and people within the Resources Agency that we need what I consider to be even greater commitment for the change of the basic function of those agencies. The basic function of those agencies should be to provide appropriate land use controls that allow man to live with the natural process of the river. First, I would suggest that the gentleman who builds an orchard in the flood plain should not be entitled to the "hand wringing and teary eyed" kind of sympathy that he generally gets from some courts. It should be pointed out that it is he who made economic decisions to plant his prune orchard where the water flows and where the river is slowly but surely going to flow one day. Second, I suggest that the benefits which will emerge from this kind of program are immense. The Corps spends hundreds of millions of dollars each year on bank protection projects and the like, and yet what do we talk about in the way of state money for the acquisition of riparian habitat? A mere two million. The Corps spends more than just working up the scratch paper studies on the beginning of a dam project. Our priorities are vastly out of line. The taxpayer could save millions of dollars and perhaps billions nationwide.

I think following the example that is being set by President Carter, a thoroughly planned analysis of the basic mission of the Army Corps of Engineers and the Bureau of Reclamation must be done. That analysis is going to come only if people in our government, elected and otherwise, have a perception that it's what the public wants. Without that dedicated and hard working political action, we're not going to get it. I hope that those people who are most interested in riparian habitat will organize better and participate more in the hearings that are going to be coming up in the future, just like the hearings we had last year. I think if they do, the result will be change. There's one thing sure and that is it won't be changed if you don't participate. Secretary of the Interior Andrus put it best when he said, "the rivers have a multitude of values other than simply irrigation and providing hydroelectric power." He said that perhaps the most important value to retain is that tenuous link between man and nature. Right now, the government agencies are almost solely dedicated to destroying that link. That has to change; it can be changed, but only with your participation.

Panel Discussion

The following questions and responses were selected to represent three broad areas which were discussed by the panel of speakers: (1) bank protection projects, (2) incentives to owners of riparian forests to encourage protection, and (3) re-establishment of riparian forest habitats.

Question: (from the audience) Suppose action were taken and channelization and bank protection projects were stopped. What would happen to the river if such projects were stopped when only half finished? Especially where channelized or riprapped areas were not contiguous?

Answer: (Daniel Frost) The river banks would slowly but surely return to their natural state. The bank protection sites would slowly erode away. One way to protect the river environment, including the riparian forests, is to stop the bank protection programs which have the "mission" of freezing the river in its existing channel. It is not enough to stop bank protection, you have to go further. My own feeling is that the best solution is to work for a coastal-type bill under which riparian habitat can not be removed without at least a state permit. It would not involve condemnation. I think that it should be coupled with an aggressive acquisition program.

Question: (Betsy Marchand, Yolo County Supervisor) Is special tax treatment such as the Williamson Act available for property owners of riparian forests who wish to preserve them?

Answer: (James Burns) Tax incentives are not great enough at this time because the orchardmen can make a lot more money converting riparian forest to walnuts, prunes and almonds. These are luxury products with a high market value. The questions then arise "what about the world's protein deficit and why not plant soybeans?" Their reply is that when California has to feed the world, _then_ we'll convert prune orchards to soybeans.

Question: (from the audience) What would be an effective strategy to allow the riparian ecosystem to evolve on a piece of land which is now in use as an orchard?

Answer: (Warren Roberts) We are going to be establishing a small riparian woodland on a relatively undisturbed site in the UCD Arboretum. One of the typical characteristics of a lot of the riparian species is that they are very fast growing. The cottonwoods, willows, box elders, and walnuts on this rich, well-watered land will grow fast. It's true that the oaks are a bit slower. A person could re-establish a very effective woodland and in less than 20 years, it would evolve into a very attractive habitat. I think that brings up another point. It seems to me that while we are thinking about preserving riparian habitat, we could also think of some of this land as productive timber land. This is extremely productive soil, and we can think of it producing five year crops on properly managed forestry programs as well as providing wildlife habitat.

Comments: (Richard Stallcup) As president of the Western Field Ornithologists (a society of professional and serious amateur ornithologists of more than 1200 members which publish the quarterly journal _Western Birds_) I speak from the philosophy of its membership.

We feel that riparian woodland is a vital part of the remaining natural environment that _is_ California. This habitat, now contained in merely a shadow of its former acreage, is unique in its exceedingly lush vegetation and in the diversity of wildlife upon which it depends. There is no habitat quite like riparian woodland in the temperate world, and its quality as an ecosystem can only be compared to the tropics. We strongly urge that the remaining stands of riparian forest, which still house animal life historically typical of this habitat, are saved at any cost.

This problem demands immediate attention! We worry about endangered species, much, we are afraid, out of the guilt we bear for our hand in their strife. The loss of an entire habitat is even a larger burden to carry, for this is the life support of numerous plants and animals, many of which are rare and endangered. If we lose the riparian, like much of the great Mangrove-Cypress swamps of the southeastern United States, we will have lost another significant part of the natural heritage of North America. Again, the shame will be ours forever.

Conclusion: Following the interactions between panelists and audience came a unanimous appeal for action to save our remaining riparian ecosystems. The challenge was accepted, and a citizens' group called the Riverlands Council developed. This organization is a coalition of landowners, biologists, fishermen, bird watchers, canoeists and kayakers, farmers, educators and other appreciators of riverlands. It will work to preserve riparian habitat through public education, legislation, fund raising and land acquisition. Send inquiries to: Riverlands Council, P. O. Box 886, Davis, California 95616.

Riparian Forest Field Trip

On May 15, 1977, John Anderson and Mark Longwood led thirty-four symposium participants on a field trip through the riparian forests of Bobelaine Sanctuary (managed by the Sacramento Audubon Society) on the Feather River. This trip was enhanced by magnificent forest jungles, blooming wild roses, fruiting grape vines, and fifty-nine species of birds. We were again reminded of the immense productivity of riparian ecosystems, and came away with renewed determination to protect the remaining vestiges of this unique California resource.

Birds Observed on Field Trip to Bobelaine

Great Blue Heron
(*Ardea herodias*)

Green Heron
(*Butorides virescens*)

Great Egret
(*Casmerodius albus*)

Black-crowned Night Heron
(*Nycticorax nycticorax*)

Mallard
(*Anas platyrhynchos*)

Wood Duck
(*Aix sponsa*)

Turkey Vulture
(*Cathartes aura*)

Red-tailed Hawk
(*Buteo jamaicensis*)

California Quail
(*Lophortyx californicus*)

Ring-necked Pheasant
(*Phasianus colchicus*)

Killdeer
(*Charadrius vociferus*)

Spotted Sandpiper
(*Actitis macularia*)

Forster's Tern
(*Sterna forsteri*)

Mourning Dove
(*Zenaida macroura*)

Great Horned Owl
(*Bubo virginianus*)

Belted Kingfisher
(*Megaceryle alcyon*)

Acorn Woodpecker
(*Melanerpes formiocivorus*)

Downy Woodpecker
(*Dendrocopos pubescens*)

Nuttall's Woodpecker
(*Dendrocopos nuttallii*)

Western Kingbird
(*Tyrannus verticalis*)

Ash-throated Flycatcher
(*Myiarchus cinerascens*)

Black Phoebe
(*Sayornis nigricans*)

Gray Flycatcher
(*Empidonax wrightii*)

Western Wood Pewee
(*Contopus sordidulus*)

Olive-sided Flycatcher
(*Nuttallornis borealis*)

Tree Swallow
(*Iridioprocne bicolor*)

Bank Swallow
(*Riparia riparia*)

Rough-winged Swallow
(*Stelgidopteryx ruficollis*)

Cliff Swallow
(*Petrochelidon pyrrhonota*)

Scrub Jay
(*Aphelocoma coerulescens*)

Yellow-billed Magpie
(*Pica nuttalli*)

Common Crow
(*Corvus brachyrhynchos*)

Plain Titmouse
(*Parus inornatus*)

Bushtit
(*Psaltriparus minimus*)

White-breasted Nuthatch
(*Sitta Carolinensis*)

Wrentit
(*Chamaea fasciata*)

House Wren
(*Troglodytes aedon*)

Bewick's Wren
(*Thryomanes bewickii*)

Mockingbird
(*Mimus polyglottos*)

American Robin
(*Turdus migratorius*)

Swainson's Thrush
(*Catharus ustulatus*)

Western Bluebird
(*Sialia mexicana*)

Northern Shrike
(*Lanius excubitor*)

Starling
(*Sturnus vulgaris*)

Hutton's Vireo
(*Vireo huttoni*)

Warbling Vireo
(*Vireo gilvus*)

Yellow Warbler
(*Dendroica petechia*)

Townsend's Warbler
(*Dendroica townsendi*)

Wilson's Warbler
(Wilsonia pusilla)

Western Meadowlark
(Sturnella neglecta)

Northern Oriole
(Icterus galbula)

Brewer's Blackbird
(Euphagus cyanocephalus)

Brown-headed Cowbird
(Molothrus ater)

Black-headed Grosbeak
(Pheucticus melanocephalus)

Lazuli Bunting
(Passerina amoena)

House Finch
(Carpodacus mexicanus)

American Goldfinch
(Spinus tristis)

Rufous-sided Towhee
(Pipilo erythrophthalmus)

Brown Towhee
(Pipilo fuscus)

Curt Acredolo, 2709 Adrian Drive, Davis, CA 95616.

Donald Alley, Dept. Wildlife and Fisheries Biology, Univ. of Calif., Davis CA 95616.

Dan Anderson, Dept. Wildlife and Fisheries, Univ. of Calif., Davis, CA 95616.

John Anderson, 3615 Auburn, Sacramento, CA 95822.

Cheryl & William Appleby, 1052 Manor, Marysville, CA 95901.

Vickie Ashmore, 317 S. 20th Avenue, San Jose, CA 95116.

Sue Ayer, 1608 Orange Lane, Davis, CA 95616.

Patrick E. Baker, 27460 Lyford St., Hayward, CA 94544.

Katie Balderson, Route 1, Box 1895, Davis, CA 95616.

Bill Barclay, 728 E. 10th St., Davis, CA 95616.

Claire Barrett, 437 F Street, Davis, CA 95616.

Cameron Barrows, Route 1, Box 1895, Davis, CA 95616.

Jim Barry, Department Parks and Rec., PO Box 2390, 1415 8th, Sacramento, CA

Michael G. Barbour, Dept. of Botany, Univ. of Calif., Davis, CA 95616.

Nathan B. Benedict, 11B Cuarto Hall, Univ. of Calif., Davis, CA 95616.

Irene Biagi, 330 Forbes Avenue, San Rafael, CA 94901.

Gary Bolton, 7981 Santa Barbara Dr., Rohnert Park, CA 94928.

John Brach, 754 Murick St., Red Bluff, CA 96080.

Claudia L. Breault, 525 Clayton, San Francisco, CA 94117.

C. Brown, DFG-PO Box 607, Red Bluff, CA 96080.

Peter Brown, 1063 Olive Drive, Davis, CA 95616.

Roland Boogthold, 1029 Audrey Way, Roseville, CA 95678.

Jim Burns, 5130 El Cemonte, Davis, CA 95616.

Pam Burton, 919 Drake, #142, Davis, CA 95616.

J.J. Cech, Dept. of Wildlife and Fisheries, Univ. of Calif., Davis, CA 95616.

Mike Center, 427 K Street, Davis, CA 95616.

Karen Chin, 5253 Countryside Dr., San Diego, CA 92115.

Beverly Clark, 2713 Bidwell, #3, Davis, CA 95616.

Kathy Clark, PO Box 1662, Auburn, CA 95603.

Illa Collin, 7423 Braeridge Way, Sacramento, CA 95831.

Marjorie Colt, 135 Redondo Drive, Pittsburg, CA 94565.

Susan Conard, Dept. of Botany, Univ. of Calif., Davis, CA 95616.

Cindi Corwin, 1025 University, #165, Sacramento, CA 95825.

Elizabeth Coss, 623 E. St., Davis, CA 95616.

Alan Craig, Dept. of Fish and Game, 3532 Winston Way, Carmichael, CA 95608.

Earle Cummings, 13206 Jackson Rd., Soughhouse, CA 95683.

David Davenport, 1714 Poplar Lane, Davis, CA 95616.

Russell Dedrick, 1311 L. St., Davis, CA 95616.

John Dentler, 716 Elmwood Drive, Davis, CA 95616.

Marisela de Santa Anna, 7811 Lynch Rd., Sebastopol, CA 95472.

Denise Devine, 1910 Salem St., Chico, CA 95926.

Senator John Dunlap, State Capital, Sacramento, CA 95828.

Donna Durham, 1409 Maplewood Dr., Modesto, CA 95350.

Elizabeth Dutzi, Dept. of Geography, Univ. of Calif., Davis, CA 95616

George Faggella, 2905 Austin #1, Davis, CA 95616.

Kathy Farley, 328 D. St., Davis, CA 95616.

Bill Felts, 3034 Bryant Place, Davis, CA 95616.

Kent Fickett, 1900 Sunshine Dr., Concord, CA 94520.

Sandra Fielden, 2004 Haring, Univ. of Calif., Davis, CA 95616.

John Fielding, DWR-PO 388, Sacramento, CA 95802.

Cheryl Folsom, 1115 Grandview, Napa, CA 94558.

Steve Ford, 424 F. St., Davis, CA 95616.

Tim Ford, 924 E. Fairmont, Modesto, CA 95350.

Catherine Fox, 1742 Curtis St., Berkeley, CA 94702.

Martin Franco, 501 E. 7th St., Davis, CA 95616.

John Franklin, 216 W. 8th St., Davis, CA 95616.

Sara Frost, 2431 El Verano, Redding, CA 96001.

Dave Gaines, Box 33, Branscomb, CA 95417.

Gerry Germain, 3971 Oak Hurst Circle, Fair Oaks, CA 95628.

Eric Gerstung, 1132 McClaren Dr., Carmichael, CA 95602.

Paul Giguere, 5307 Marconi, Carmichael, CA 95602.

Thomas Graham, 2400 Davison, Davis, CA 95616.

John Gray, 120 C Street, #5, Davis, CA 95616.

Randall Gray, 4260 Silver Cress Ave., Sacramento, CA 95821.

Joanne Hagopian, c/o Vic Fazio's Office, State Capital Bldg. Rm. 5126, Sacramento, CA 95828.

Don & Sharon Hallberg, 1408 Drexel Dr., Davis, CA 95616.

Lori Hamburger, 801 D St., #4, Davis, CA 95616.

A. Hamilton, 1860 Church St., San Francisco, CA 94131.

W. Edward Harper, 5151 Greenberry Dr., Sacramento, CA 95841.

Stephen P. Hayes, 1803 Poplar Ln., Davis, CA 95616.

L.R. Henrich, Rt. 1—Box 152, Glenn, CA 95943.

Ralph Hinton, DWR, PO 687, Red Bluff, CA 96080.

Robert F. Holland, Agronomy Dept., Univ. of Calif., Davis, CA 95616.

Leslie Hood, CNACC, 1505 Sobre Vista, Sonoma, CA 95475.

Keith Hopper, Div. of Env. Studies, Univ. of Calif., Davis, CA 95616.

J.L. Hornbeck, 609 Anderson, Davis, CA 95616.

Paul Howard, 555 Audubon Place, Sacramento, CA 95825.

Greg Howe, 1822 Apple Ln., Davis, CA 95616.

Richard Hudson, 1333 Locust Pl., Davis, CA 95616.

Robert J. Hird, 1422 Drake Dr., A, Davis, CA 95616.

J.K. Inuzuka, Rt. 1, Box 22, Winters, CA 95694.

Carl Jochums, 923 Fulton Ave., #425, Sacramento, CA 95925.

Bruce Jones, 5220 M Street, Sacramento, CA 95819.

Sally Judy, Box 33, Branscomb, California 95417.

Dean & Sally Jue, 1111 J St., #87, Davis, CA 95616.

B. Kay, Agronomy Dept., Univ. of Calif., Davis, CA 95616.

Ed Keller, Div. of Env. Studies, Univ. of Calif., Santa Barbara, CA 93106.

Dave Kelly, 129 Hedy Lane, Davis, CA 95616.

Ken Kemmerling, PO 558, Davis, CA 95616.

Fred Kindel, 555 Douglas, Broderick, CA 95605.

Dick Kroger, 2800 Cottage Way, Rm. 2727, Sacramento, CA 95825.

Paul Kryloff, 330 Forbes Ave., San Rafael, CA 94901.

Rebekah Leiser, 2644 Belmont, Davis, CA 95616.

Gail Lemlly, 1205 Stanford Pl., Davis, CA 95616.

Chris Lewand, 571 Rainsville Rd., Petaluma, CA 94952.

Kelly Logan, PO 14, Camp Meeker, CA 95417.

Marc Longwood, Biological Science, Calif. State Univ. at Sacramento, Sacramento, CA 95819.

Rod MacDonald, Dept. of Botany, Univ. of Calif., Davis, CA 95616.

Gary Macey, 1107 5th St., Davis, CA 95616.

Jack Major, Dept. of Botany, Univ. of Calif., Davis, CA 95616.

Robert Martin, 338 Bridge Pl., West Sacramento, CA 95691.

Wendy Matyas, PO 56, North San Juan, CA 95950.

Katherine McBride, 1545 Drake Drive, Davis, CA 95616.

Sandi McCubbin, c/o Senator Dunlap, State Capital, Sacramento, CA 95828.

Robert R. McGill, PO 607, Red Bluff, CA 96080.

J.L. Medeiros, 1409 Maplewood Dr., Modesto, CA 95350.

J. Melo, 119 Jewett St., Ft. Bragg, CA 95437.

Ellen Miller, 1918 California Rd., Modesto, CA 95351.

Mike Miller, 2900 Cottage Way, Rm. E 2727, Sacramento, CA 95823.

Karen Miller, 2321 P St., Sacramento, CA 95816.

Marcie Moloshco, 541 Rainsville Rd., Petaluma, CA 94952.

Jeff Moncur, 955 Cranbrook St., #300, Davis, CA 95616.

Bob Motroni, 3465 Rattlesnake Rd., Newcastle, CA 95658.

Peter Moyle, Dept. of Wildlife and Fisheries, Univ. of Calif., Davis, CA 95616.

Pam Muick, c/o Cotati Coop, Old Redwood Highway, Cotati, CA 94928

Claude Nagamine, Land, Air and Water Resources, Univ. of Calif., Davis, CA 95616.

Jim Nelson, 131 Camp Joy Rd., Boulder Creek, CA 95006.

Barbara Nieb, 10940 E. Hainey Ln., Lodi, CA 95240.

Sean O'Connell, 210 Prospect Dr., San Rafael, CA 94901.

Tedmund Oda, PO 496, Walnut Grove, CA 95690.

Glenn Olson, 555 Audubon Place, Sacramento, CA 95825.

Mark Otto, 228 D St., Davis, CA 95616.

Noreen Parks, PO 725, Twain Harte, CA

Bill Patterson, 5323 Jerome Way, Sacramento, CA 95819.

Bruce Pavlik, 619 Lesslet Pl., Davis, CA 95616.

David Peart, 2T Solano Park, Davis, CA 95616.

Charlie Pike, 1416 Ninth, Rm. 252-26, Sacramento, CA 95814.

Dave Pratt, 628 Barbera Pl., Davis, CA 95616.

Greg Probst, 619 Pole Line Rd., #133, Davis, CA 95616.

Larry Puckett, 1510 El Cerrito Dr., Red Bluff, CA 95080.

Crispin Rendon, 3968 3rd Ave., Sacramento, CA 95817.

Marlene Rhoda, 11E Solano Park, Davis, CA 95616.

David Rhode, 114 W. 8th St., Davis, CA 95616.

John & Mary Rieger, 4123½ Wilson Ave., San Diego, CA 92104.

Warren G. Roberts, 2223 Amar Ct., Davis, CA 95616.

Robert Robichaux, 620 D. St., Davis, CA 95616.

Ken Roby, USFS, Fish & Wildlife, 630 Sansome St., San Francisco, CA 94111.

Mathew Rowe, 1443 Wake Forest Dr., #2, Davis, CA 95616.

Anne & Pete Sands, Rt. 1, Box 2230, Davis, CA 95616.

Mike Sadleir, 1605 Arden Dr., Woodland, CA 95695.

Jim Scarff, 133 17th St., #7, Oakland, CA 94612.

Phil & Margaret Schaeffer, 376 Greenwood Beach Rd., Tiburon, CA 94920.

Bob Sculley, 426 E. 8th St., #1, Davis, CA 95616.

G. Serr, PO 607, Red Bluff, CA 96080.

Ahmed Sidalmed, 1850 Hanover Dr., #150, Davis, CA 95616.

Michael Singer, Land, Air and Water Resources, Univ. of Calif., Davis, CA 95616.

Diane Siskey, 4629 Marconi, #107, Sacramento, CA 95821.

C. Sluaeder, PO 621, Broderick, CA 95605.

Felix E. Smith, 2800 Cottage, Sacramento, CA 95825.

Evan Snyder, Dept. of Entomology, Univ. of Calif., Davis, CA 95616.

Richard Soehren, 2400 Q, #32, Sacramento, CA 95816.

Marian Stephenson, 507 E. 8th St., Davis, CA 95616.

W. Stiles, 1919 Fruitale Ave., J624, San Jose, CA 95128.

Tom Stone, Box 1480, Redding, CA 96001.

Dean Taylor, Dept. of Botany, Univ. of Calif., Davis, CA 95616.

Kathy Terry, Coastal Commission, 701 Ocean St., Rm. 310, Santa Cruz, CA 95066.

Lynn & Syd Thomas, Rt. 2, Box 188, Chico, CA 95926.

Susi Thompson, 2412 Cottage Way, Sacramento, CA 95825.

Kenneth Thompson, Dept. of Geography, Univ. of Calif., Davis, CA 95616.

Gene Trapp, Rt. 1, Box 1155, Davis, CA 95616.

Cal Watch, Box 621, Broderick, CA 95605.

Doug Weinrick, 7617 Scoan Tres Blvd., San Jose, CA 94078.

Bradlee Welton, 401 San Miguel Way, Sacramento, CA

Alan Whittemore

Jack Wilburn, 2113 Bueno Drive, Davis, CA 95616.

Patty Willcocks, 703 6th St., Davis, CA 95616.

Stephanie Williams, 2626 Belmont Dr., Davis, CA 95616.

Roger Wirllmarth, 1107 9th St., #419, Sacramento, CA 95814.

Dave Winkler, 423 E St., Davis, CA 95616.

Jeffrey Wood, 4817 Quonset Dr., Sacramento, CA 95820.

Chet Zajac, 3968 3rd Ave., Sacramento, CA 95128.

Mary Burke, 623 E Street, Davis, CA 95616.